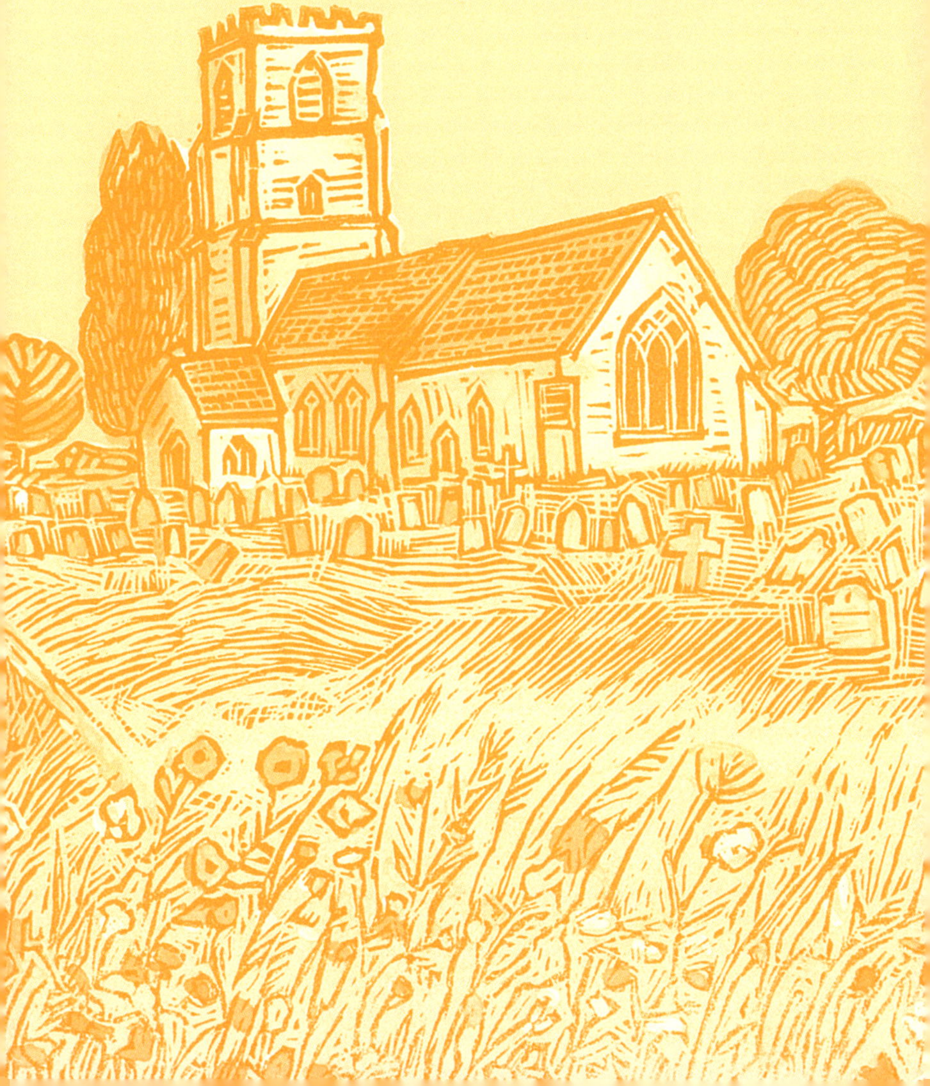

WHERE ARE THE FELLOWS WHO CUT THE HAY?

With special thanks to Cherry Tree Farm and
Rosedale Funeral Home for their generous
support as patrons of this book

WHERE ARE THE FELLOWS WHO CUT THE HAY?

*How Traditions From the Past
Can Shape Our Future*

ROBERT ASHTON

unbound

First published in 2024

Unbound
c/o TC Group, 6th Floor Kings House, 9–10 Haymarket,
London, United Kingdom, SW1Y 4BP
www.unbound.com
All rights reserved

© Robert Ashton, 2024
Illustrations © John Richardson, 2024
Foreword © Paul Heiney, 2024

The right of Robert Ashton to be identified as the author of this work has been asserted in accordance with Section 77 of the Copyright, Designs and Patents Act, 1988. No part of this publication may be copied, reproduced, stored in a retrieval system or transmitted in any form or by any means without the prior permission of the publisher, nor be otherwise circulated in any form of binding or cover other than that in which it is published and without a similar condition being imposed on the subsequent purchaser.

Typeset by Jouve (UK), Milton Keynes

A CIP record for this book is available from the British Library

ISBN 978-1-80018-298-1 (hardback)
ISBN 978-1-80018-299-8 (ebook)

Printed in Great Britain by Clays Ltd, Elcograf S.p.A.

To my father-in-law Michael, who taught me how to plough a straight furrow.

Contents

Foreword by Paul Heiney	ix
Introduction: Silence	xi
LIFE	1
Milk	5
Wheat	22
Wool	39
WORK	55
Leather	59
Barley	76
Coins	94
POWER	109
Coal	113
Steam	128
Iron	140
COMMUNITY	157
School	161
Church	175
Trade	189
Afterword	205
Acknowledgements	207
References	209
Supporters	217

Foreword

I STARTED A SMALL farm in Suffolk in the early 1990s with the idea of recreating the workings of a modest Victorian farm. My passion was the Suffolk Punch, the great carthorse of the eastern counties, and so I employed these massive equine powerhouses rather than spluttering tractors. I had learned a little horsemanship, and been shown the workings of the plough and all the other farming machinery that was needed to cultivate the land.

But one thing was lacking: experience. Now, had I been a modern farmer it would have been no problem to turn to my neighbours for advice, but did any of them understand the workings of the horse, or even remember the way things were in the days I was trying to recreate? Of course not, and in my quest for experience I began to realise, as the writer George Ewart Evans had done some decades before, that a strong tide was flowing and taking with it a way of life that had endured for centuries. So I read George's book and lapped up every page, picking out useful bits of instruction from his narrative.

In this book, Robert Ashton has done something similar. Like me a great fan of George Ewart Evans, he has used his writings to place into context some of his memories of working the land and of rural life as it was lived fifty or more years ago. This book, alongside Evans's, is not simply an exercise in nostalgia. There was much wrong with 'the old ways of working': often unsafe, unhealthy and poorly paid. But there was a lot that was right about it when it came to a compassionate relationship with farm animals and a respect for the delicate balances within the soil that we have discarded in pursuit of commerce. And that, I believe, is the lesson of this book: look back not with misty eyes, but with an enhanced appreciation of a gentler, more sustainable way of working the land, and the dedication of the people who worked it.

Paul Heiney, Suffolk 2023

Introduction

Silence

WHERE ARE THE FELLOWS who cut the hay, I asked myself as I sat one Sunday morning in the silence of Leiston Quaker Meeting? Farming practices and even the landscape itself have changed so much in the fifty years since I first worked on the land. Fields are larger, tractors are larger and far fewer people now are needed to grow crops, milk cows and feed pigs. I thought back to the companionable days I'd spent in my youth, hoeing sugar beet as part of a gang, walking up and down the field in a line, smiling at the cursing of a workmate when he discovered that his row of beet contained more weeds than everyone else's. I thought back to the summer of 1969, when I turned hay and later baled it with a Fordson Super Dexta, a tiny tractor by today's standard. What happened to these people and those who came before them, and to their way of life?

These thoughts came to me that morning in autumn 2016, because this was my first visit to this particular Quaker Meeting and I'd realised that Florence Evans must have once sat in this very room, in that contemplative, silent space that we Quakers call worship. She had lived in the nearby village of Blaxhall in the early 1950s and this had been the closest Quaker Meeting to her home. Later, in the early 1960s, she had been my school headmistress in Needham Market and, as I discovered years later, was the wife of George Ewart Evans, a writer and oral historian whose work had so inspired me in my teens and in whose footsteps I hoped to tentatively tread when I could find the time to make writing my full-time occupation.

George Ewart Evans had interviewed his neighbours in Blaxhall, writing down the stories they told him of life before the First World War. He spoke with men who mourned the change from horses to tractors and men who had worked at Snape Maltings decades before Benjamin Britten saved it from dereliction and created a world-class concert hall. He had not set out to become to spoken history what Cecil Sharp was to folk music, but, chatting over the garden fence with his neighbours Robert and Priscilla Savage, he realised that unless someone wrote down their stories, and the stories of others who could remember how things used to be, they would be lost forever.

When the Evanses moved to Blaxhall in 1948, Robert Savage was sixty-eight and retired, after having been a shepherd all his life. The Savages became family friends, and not surprisingly featured prominently in George's book *Ask the Fellows Who Cut the Hay*, which was first published in 1956. This was the book that established his reputation as a collector of stories from our rural past. Evans wrote not in a nostalgic or romantic way, but in a more matter-of-fact way, because he knew that time never stands still, of how rural life was changing.

Introduction

The year before my visit to Leiston Quaker Meeting, by chance my wife Belinda had spotted that young Jessica Juby, who had for a time worked as my assistant, was selling a collection of Evans's books. Her grandfather, Alan Juby, had lived at Reymerston in Norfolk and over a long lifetime built up quite a collection of books about rural life. His brother had kept heavy horses and Alan had driven a lorry for Roger Warnes, a farmer and haulier to whom in my own early career I had sold fertiliser. Belinda bought the books for me for my sixtieth birthday that August.

Reading them strengthened my resolve to write about rural Suffolk. How had things changed since Evans's day? Would people in their eighties today talk in similar ways to those Evans had interviewed, about the changes they had witnessed over their lifetime? Would they say that although life had become easier, something had been lost, just as those Evans interviewed had told him? And how was the farming world responding to the growing concern about climate change? Was the race to produce more being tempered by a more pragmatic focus on sustainable profitability rather than yield alone?

My attention snapped back to the room in which I was sitting, as the hour of silent worship had come to an end and, as is the tradition, people were shaking hands with those sitting around them. I had not long been a Quaker then, and when as a teenager I'd lived a couple of hundred yards from the Meeting House I'd not dreamed of ever crossing the threshold. But we all change as we journey through life, and I had increasingly felt drawn to the Religious Society of Friends, and became a member by what Quakers call 'convincement'. This is the term used to describe when someone finds Quaker values and practices congenial enough to apply for membership.

Quakers have a rich history of blending entrepreneurship with social action, and this, together with their more liberal theology, had attracted me. Florence Evans was what is known as a birthright Quaker, born to Quaker parents, and remained in membership throughout her life. Her children had attended the Quaker school at Saffron Walden, and I think that throughout her working life she was a Quaker first and a teacher second.

I was in my mid-teens when one Christmas my parents gave me a copy of *Ask the Fellows Who Cut the Hay*. It is described by its publisher, Faber, as 'a vivid portrait of the rural past of Blaxhall, a remote Suffolk village, in the time before mechanization changed the entire nature of farming, the landscape and rural life for good'. At the time I read the book I was working on farms just a few miles from Blaxhall, and found so much of what Evans had written familiar. I found a resonance between the stories he relayed and my own experience, and discovered that I was working alongside family members of those Evans had interviewed.

Blaxhall is just seven miles from Leiston and the Meeting House, and the farms I had worked on all lay between the two villages. George Ewart Evans had also been a teacher, but increasing deafness forced him to give up his career, and so while Florence was at work in Blaxhall's school, George was in the schoolhouse looking after their children and, when he could, writing. Today he might be described as a house husband, but back then it was unusual for the woman of the house to be the main breadwinner. This perhaps did little for his self-esteem, but his interest in the past, a natural curiosity and a wish to be a good neighbour led to conversations over the garden fence that grew to become his life's work as an oral historian.

Paradoxically, his deafness made him a good listener and he wrote down the stories those he met in the village told; stories of a way of life that appeared to have changed little down the centuries. He encouraged people to talk about the past, sharing their memories of times that were fast being forgotten. He was not new to collecting the memories shared by those who lived around him. As a young man during the 1926 General Strike, he had walked with out-of-work miners in the hills above Abercynon, the village where he was born and grew up. I suspect that in both Wales and Suffolk people were more than happy to talk to someone who showed a keen interest in what to them may have seemed unimportant tales from their pasts.

Evans had read classics at university, and as a historian noticed that many of the dialect words his Blaxhall neighbours used had remained unchanged since Chaucer's day. Later, he recorded many of his interviews, capturing the voices of village men and women as they spoke fondly of what was rapidly becoming a bygone age. He would set his tape machine running, ask one or two open questions, and let them take the conversation where they wanted it to go. The stories he heard were documented in his books.

I married into a local farming family and that autumn morning left my wife visiting her aged parents while I attended the Quaker Meeting in nearby Leiston. My father-in-law, together with neighbouring farmers who were family friends, had been working the land at the time Evans was living at Blaxhall and writing his book. They too talked about the change from horse to small tractor, then to the huge tractors of today; tractors that pull giant ploughs and require hedgerows to be removed so as to create enormous prairie-like fields they can till quickly and efficiently. The very landscape itself had now changed, as well had the lives of those who live and work upon it.

It struck me that day, sitting in Leiston Quaker Meeting, that so much of what I now know and cherish will also have been familiar to Florence Evans. She too will have been familiar with the quaint language and traditions that set Quakers apart from other religious groups. Her husband George was not a Quaker, being I think tolerant though sceptical of his partner's faith. This is also the case in my own household, where my wife feels no need to spend her Sunday mornings in the company of others, searching for spiritual meaning. My visit to

Leiston Quaker Meeting gave me a new sense of connection with Florence Evans and made me even more determined one day to write a book that builds on George's work.

When I had moved to Needham Market as a small boy and Florence Evans had been my school headmistress, I had no idea that her husband was a writer. Nor did I know that in that same year, 1961, *Ask the Fellows* had been so well received that it was reprinted and remains in print today. To me Florence was just a nice, friendly headmistress in a school that I, as a rather shy newcomer, was finding rather daunting.

Needham Market had a Quaker Meeting House and was where local Quaker Samuel Alexander had opened Alexander's Bank in 1744. This and other local banks, including Gurney's of Norwich, had merged in 1896 to form Barclays Bank. Samuel Alexander, Robert Barclay and Samuel Gurney had also been Quakers. That was the bank branch where my father had come to work. I suspect Florence felt at home being a Quaker in Needham Market.

The primary school, just off Needham Market High Street, had not been a Quaker school. It was situated on a lane called The Causeway, which in this instance was said to be a corruption of 'the corpseway', as it was the isolated route along which plague victims had been carried for burial at nearby Barking church. This historic fact used to fascinate me and I would often walk out of the village, past where The Causeway became a track, just as it had all once been, trying to imagine what those funeral processions would have been like. I had a very fertile imagination.

The school was a short walk from the town's Quaker Meeting House and the values by which Florence lived her life certainly rubbed off on the children she taught. She was a cheerful woman with her grey hair tightly netted and a small bun at the back. I cannot remember any other teachers from those days. I suspect Florence's calm nature and conviction that she was working for God as well as the local education authority must have made her more memorable than the classroom teachers.

She was never one to raise her voice or lose her temper. I recall one term when we had some visiting Gypsy children in my class. I found them interesting and would play with them. I have always liked people who stand out from the crowd. Other children, however, were less enthusiastic, perhaps having picked up their parents' prejudices. Florence made every effort to include these outsiders in the life of the school. She would set us tasks, such as finding a particular flower or leaves from the tree in the playground. We'd be sent in twos, with the Gypsy children paired with those she thought would benefit most from getting to know them better. Decades later, I realised that this was probably prompted by the Quaker testimony to equality. It was also one of those pivotal episodes in my early life, when it became inevitable that I too would one day become a Quaker.

When I was thirteen, we left Needham Market. My father was appointed manager at Barclays Bank in Leiston, a small rural town with a rich industrial history. It was here in 1852 that Richard Garrett established Britain's first purpose-built production line, making steam engines. It was these engines that began the mechanisation of farming and the changes to rural life that George Ewart Evans later wrote about. At its peak Garretts employed 2,500 men at its factory, which was in the middle of the town, opposite Barclays Bank. More than half of its production was exported. By the end of the nineteenth century, Garretts of Leiston was well known all round the world. Today the Long Shop, which housed the production line, is preserved as a museum telling the story of Garretts of Leiston.

In the late 1960s it was customary for Barclays' managers to live in the flat above the branch. This was no hardship in Leiston, as the bank occupied the front half of a rather grand early Victorian house which had been built by one of the Garrett family, so we had a kitchen and dining room to the rear of the bank, and four bedrooms and a large lounge upstairs. The house had extensive cellars and attics that had once been servants' quarters. My parents never ventured into the attics, so they quickly became places where my brother and I would play. Compared to the bungalow we had left behind in Needham Market, this house was huge.

Behind the bank was the garden, which must have been around an acre in size. It was surrounded by a high brick wall, the other sides of which were the streets of Victorian terraced houses built to accommodate the Garrett workers. In the middle of the garden there was a large greenhouse, and to one side a barn where the carriages had once been kept. This was where my father kept his Austin Cambridge. A wooden mission church had replaced the stables and was accessed from the road outside. It was a few feet from our kitchen window.

Singing from the mission church was not the only religious activity we overheard. Mr Smith, a local smallholder, was an enthusiastic member of the Plymouth Brethren. Every Saturday he would stand in the square outside the bank and, waving the Bible grasped firmly in his hand, loudly encouraged all who passed by to change their ways and turn to Christ. I would watch and hear him from my bedroom window, which overlooked the square. He had a very loud voice.

My father did not really approve of Mr Smith, preferring the conventionality of Anglican worship at St Margaret's Church. My parents were regular churchgoers and my father for a time a churchwarden. This he felt was an appropriate role for a bank manager. Status was everything to my father and providing everything appeared to others to be OK, then to him it was OK too. Looking back now, I can see that he went to great lengths to maintain a facade of respectability, which masked his insecurities. He would spend hours on his own, playing Beethoven's 'Moonlight Sonata' on his piano, with the notes becoming increasingly slurred as he consoled himself with whisky, which inevitably interfered with his fingering on the keyboard.

He sent my brother and me to confirmation classes in the vicarage and, later, I became an altar server at the church. This was something he could talk about, to his customers, colleagues and those seemingly clandestine Lodge friends, for whom he played the organ at their masonic ceremonies. My brother, perhaps because he was younger than me, seemed to quite successfully avoid going anywhere near the church after his confirmation. Our sister was sent to board at a convent school at Ditchingham near Bungay and religion remained part of her life into adulthood.

Just along the road from our home above the bank is the crossroads, where the Quaker Meeting House stands. On the opposite corner is the White Horse Hotel. It was the only place in town where my father felt comfortable going for a drink. He was something of a snob, and to drink in one of the local pubs risked bumping into his customers, whom he preferred to speak to by appointment, in his office at the bank. Often in the small bar of the White Horse there would be just my father and Len the hotel proprietor, a small man with little hair and no evident sense of humour. He and my father would spend evenings together being miserable, with my father drinking and Len slowly profiting from each whisky he poured.

Leiston had seen better days when we arrived in 1968. In the nineteenth century, it had been a booming, prosperous town, but now it was overshadowed by nearby Aldeburgh. The White Horse Hotel had been a coaching inn, and was a convenient place to accommodate the many visitors the success of Garretts was bringing to the town. The railway arrived in the town in 1860, with a spur running from the station down to the Garretts town-centre works. Later it was extended to nearby Aldeburgh, a town that more recently increased in popularity as rapidly as Leiston's prosperity declined. The Quaker Meeting House was also built at this time, along with many new houses to accommodate the growing workforce.

By the 1960s when we arrived, Garretts were making dry-cleaning and cardboard-box-making machines. Neither were world-class, as the steam engines had been, and while my father was their bank manager Garretts went bust, was sold and the works closed, with many men losing their jobs and pensions. Even the recently built nearby nuclear power station did little to revitalise the local economy, never employing many more than 500 people. To this day, Leiston remains a little down at heel, despite being situated between the popular seaside towns of Southwold and Aldeburgh. Affluence bypassed the town.

I must have walked past the Leiston Quaker Meeting House a thousand times as a teenager. Never once did I think that one day I would go there as a Quaker on a Sunday morning. But that is where I found myself that day, remembering Florence Evans and her husband George. But how did a bank manager's son become so interested in farming and rural village life? To explain this I need to take you back into to the garden behind Barclays Bank – because that was where it all began.

We'd not lived in Leiston long when I decided I wanted to keep chickens. I liked eggs and my mother made a lot of cakes. She said she'd pay for any eggs my chickens laid, so I did my research and my father agreed to buy me a chicken shed and some hens for my fourteenth birthday. He acquired the shed from a customer who had a smallholding on the edge of the town. It was on iron wheels, and I spent a summer holiday cleaning, repairing and creosoting it.

The only problem was how to get the shed from the smallholding to our garden. My father asked Michael, one of his local farmer customers, if he could help. This he kindly did, and one Saturday morning he fetched it with his tractor, then helped me position it at the far end of the walled garden. Seeing my enthusiasm, he asked if I'd like a weekend job on his farm. He kept pigs and a few suckler cows, and grew sugar beet, wheat and barley. There was always something to do.

Michael then was just thirty years old, although he'd had his own farm since his late teens, so was confident in ways my father could never have been. His farm was a couple of miles away at Knodishall, and his father David's farm a little closer, on the edge of Leiston itself. Michael's grandfather William had moved to the area from Cambridgeshire, where he had left his own father's farm to work as an agricultural contractor, which in the 1920s meant visiting local farms with a Garrett steam tractor and threshing machine. Michael had farming in his blood and was comfortable with his heritage, while my father's father had worked as a printer's compositor – working-class roots I think he was keen to forget.

I leaped at the chance of a weekend job and soon became a regular face on the farm. Over the years I learned to drive a tractor, to plough, to hoe sugar beet and to muck out the pigs. I basked in the simplicity of life on the farm and the connection with the natural world that gave me. Yes, my father could shuffle paper and lend the bank's money, but on the farm success came only when you worked with nature. There was an authenticity and simplicity to working the land that I found satisfying and so much more appealing than banking, which I think was the career my father hoped my brother and I would follow. Simplicity is also another of those Quaker values I think I learned from Florence Evans when I was at school in Needham Market.

Often I would ride with Michael in his grey Bedford van when he went to the hotels in Aldeburgh to collect dustbins full of swill, the scraps that we would feed to the pigs he was fattening on contract for Wall's. The van had sliding front doors and a three-speed column gear-change. There was only a driver's seat, so any passengers sat on a grain sack placed upon an upturned wooden crate. In those days, seat belts were fitted but rarely worn.

The pigs were fattened in an open yard between the barn and other outbuildings. Pigs will eat anything, and noisily enjoyed the swill we collected from local hotels and schools each morning. At each stop, full dustbins would be replaced with scrubbed-out empty ones from the day before. They were battered, metal bins and quite heavy when full.

We would unload the van and tip the swill straight over the wall onto the concrete apron that edged the pig yard. We were supposed to boil it first, but rarely bothered. In fact, the boiler was only lit when a Wall's inspector was calling. The pigs would start screaming excitedly and running around as soon as they heard the van approaching. I don't think a crowded football stadium can generate the same volume of noise as sixty hungry pigs when they know they're about to be fed. Watching them jostling and devouring the scrapings from hotel guests' plates was always interesting. Pigs are intelligent and I think would listen out for the van and be in position by the wall even before it pulled into the farmyard.

As I was the bank manager's son, I was always invited into the house for lunch on the days I was at the farm. All of the knives, forks and spoons on the table bore the Trusthouse Forte logo. They had been salvaged from the pigswell and recycled. Nothing went to waste in those days. Other people had to pay to eat their food with hotel cutlery; we did it every day, without ever paying a penny. Farmers are always frugal, preferring not to waste anything, so using discarded cutlery was as logical as sitting on an upturned crate in the van rather than buying a van with a passenger seat.

Later, when I passed my driving test, I was allowed to collect the swill on my own. This became a weekly Saturday adventure. I loved driving the old van to Aldeburgh on a summer's morning with both front doors open. If I floored the accelerator, it might just reach seventy, but only when going downhill. There aren't many hills in Suffolk!

Soon I was working on other farms too. With Mike and Des Newson, identical twins who farmed behind the church at Aldringham, and also on Blackheath Estate at Park Farm,

Friston, where I learned to milk Jersey cows. Unlike Michael's pigs, which were clamorous and boisterous, focused only on their stomachs, I found the cows to be far gentler and more capable of showing affection.

I would cycle over to help Fred the cowman, and after a few months was trusted to milk the cows alone when Fred had his day off. I came to learn the pecking order which dictated which cows would appear in the cowshed first and over time realised that each also had a different personality. One in particular I'd had to hand milk as she had torn a teat on barbed wire. I'd do this gently, and then apply antibiotic cream to the wound afterwards to help it heal. I like to think she remembered this kindness and would come over to be stroked whenever I was out in the fields, calling the cows in to be milked.

The cowman on Blackheath Estate was the first farm worker I really got to know. Unlike Michael, he was not his own boss, but one of the many men working on a large estate. I remember being surprised by how readily he deferred to the farm manager, from whom he would accept orders without question. Getting to know Fred gave me an insight into the lifestyle and attitudes of people like those that George Ewart Evans wrote about in *Ask the Fellows* and his other books.

Fred and his wife and daughter occupied a tied cottage by the road at the top of the farm drive. In the cottage next door lived Clive, a younger man, who drove a tractor and sometimes also helped with the cows. Later Clive worked for Henry Greenfield at Leiston, and then for himself, with a smallholding at Common Farm. This was as close as most ever got to farming in their own right as land then, as now, was prohibitively expensive and so out of reach.

Fred was thin, but strong in a sinewy way. His life was shaped by his diabetes, so he would break off from milking to eat a couple of biscuits if he felt his blood sugar level was becoming dangerously low. His wife was as fat as Fred was thin – not obese, just well-rounded and jolly. They had a daughter, Cindy, and Fred also had a lady friend in Ipswich he visited each week on his one day away from the farm. Being a teenager, I used to assume that Fred was enjoying a wild romance on his days off that his wife tolerated to keep the family together, but I may have been completely mistaken. But I can still see him in my mind's eye, in his pale-blue Austin A40 Farina, freshly shaved and wearing his blue beret, nosing the car out of the farmyard gate down the road towards Ipswich.

Most of the farmworkers had been there for generations. Sons almost always joined their fathers on the estate and daughters either entered service in the big house or worked casually on the farms until they married. Nobody moved far and it came as no surprise when reading *Ask the Fellows* to learn that of the 250 people living in Blaxhall in the 1950s, one in five had the family name Ling. There were Lings working on Blackheath Estate too, who I am sure were related as Blaxhall was just five miles away.

It was while working at Blackheath Estate that I discovered George Ewart Evans. *Ask the Fellows Who Cut the Hay* was the first of his books that I read. It had been written perhaps fifteen years earlier and, because I knew the area so well, really captivated me.

So much of what I had experienced fell into place when I read his books. It gave me a sense of belonging, and perhaps context, which I think had previously been missing in my life. I was after all a bank manager's son, with all the opportunities middle-class life has to offer. Those who lived and worked on the farms seemed to be content to have little, travel little and expected nothing more than that things would continue as they were. But things were changing and, in many ways, what in the 1950s had been a gentle evolution, by the 1970s had become a rapid transformation that could never be reversed.

When I turned nineteen, I went to agricultural college and

then spent ten years selling fertiliser to farmers. Here again I came to know people who lived very similar lives to those I knew in Suffolk. Norfolk, where I first worked, also had a mix of large country estates and small family farms. In those days, the agricultural merchants I worked with were mostly family firms, but over my decade in the agricultural supply industry those small companies were gradually replaced by international grain-trading operations, with their own ships. Later in my career I sold mobile homes, and then following redundancy spend thirty years working for myself, helping charities and social enterprises grow their income.

I still visit Michael, the farmer who brought my chicken shed over and gave me a weekend job. He has been my father-in-law now for more than forty years. He retired a few years ago and the farm he rented is now cropped by contractors, working for the estate of which it always formed a part. I became part of the rural Suffolk landscape that Evans wrote about so well. I realised that morning, sitting in the silence of the Quaker Meeting House, that as my own retirement approaches, it is perhaps time to revisit rural Suffolk and my past, and explore how our past might help us shape our future.

LIFE

Here is a cycle to life as it passes down the generations: we are conceived, born, raised, work, marry, have families, grow old and die. The natural world around us remains a constant to be discovered afresh as each new generation starts to explore the world beyond the childhood home.

We start our lives drinking milk, wheat makes our daily bread and wool keeps us warm. These things have not changed over the centuries, but our relationship with them has. The men and women George Ewart Evans interviewed had an intimate connection with these basics, but today we simply buy them often without a thought as to where they come from.

The people Evans wrote about drank milk that came from cows they saw grazing in the fields around their village of Blaxhall. They knew the shepherds who tended the flocks of sheep that provided the wool their mothers, sisters and wives knitted into the clothes they wore. And as children, they had gleaned the fields, picking up ears of wheat left behind after the harvest. The village miller turned these into flour that was baked into bread in every cottage.

Today we buy our milk and bread from the supermarket and, as often as not, the clothes we wear are made from man-made fibres, woven and sewn in countries thousands of miles from our own. Will this continue to be the way? Many rediscovered the joy of baking bread during the 2020 pandemic and climate change is prompting us to return to wearing natural fibres, rather than those derived from oil and gas.

Change is a constant in our lives and always has been, but what can we take from the experience of those who lived before us, for whom community meant the people who lived around them, rather than those they connect with online?

Milk

ONE THING THAT SETS us apart from other mammals is that once weaned, many of us continue to drink milk throughout our lives. Obviously, our mothers can only feed us for so long, which is why people have been keeping cows for thousands of years. When milking machines were developed in the early years of the twentieth century it became feasible to keep cows in large numbers on farms. Until then, dairy herds were small, perhaps fewer than a dozen cows, because hand milking each cow in turn took time. In the UK today there are at any one time nearly 2 million dairy cows. Cattle were brought to Britain by the Normans and selective breeding over the centuries has reached a point where a dairy cow today will typically produce 7,900 litres of milk each year. To put that into perspective, one of those articulated petrol tankers you see on the road carries the equivalent of four and half cows' annual milk production.

My interest in milk strengthened in the late 1960s, when I took a Saturday job milking cows at Park Farm, Friston, which forms part of Blackheath Estate, a vast tract of land that stretches along the north bank of the River Alde on the Suffolk coast. It reaches from Aldeburgh to Snape, and was enlarged in 1885 when the landowner Thomas Vernon-Wentworth bought the adjoining Friston Estate.

While most dairy herds at that time were black-and-white Friesians, a high-yielding breed first imported from Holland in the 1880s, Blackheath Estate kept a herd of Jersey cows. Jerseys are best known for producing milk high in butterfat. They are smaller than other breeds, but literally punch above their weight, having a reputation among stockmen for being feisty and quick to lash out with a hind leg when annoyed. But my experience was that, if treated well, they were quite docile and sometimes even affectionate, although I did get kicked a few times, particularly if a cow did not know that I was about to attach the milking machine.

At Park Farm, the cows were tethered in the cowshed, which held thirty cows at a time, in two rows either side of a central walkway. An airline ran above the cows' heads, and it was into this that we plugged the rubber hose from the milking machine. When attached to the cow, the pulsating cups would extract the milk, which accumulated in the round stainless-steel pail that formed the base. When each cow had been milked, we emptied the pail into the churns lined up against the wall. A funnel with a paper filter was used to catch any impurities that had found their way into the milk.

When full, the churns were rolled to the dairy at the end of the milking shed, lifted onto the cooler, and a rotating stainless-steel paddle, through which ice-cold water would circulate, was placed into the churn to lower the temperature. When cooled, the churns were rolled outside and lifted onto a concrete slab, the same height as the bed of the lorry that came each day to collect them. There was a knack to lifting the churns, which contained ten gallons of milk and when full weighed more than a hundredweight. The secret was to lift, then push with one knee to swing the churn up onto the loading platform. I was younger and taller than the cowman Fred and so found this easier than he did.

Looking back, I realise that there were no checks made of the temperature to which the milk was cooled, and in summer, the churns would sit outside in the full evening sun, then overnight until the lorry called at around ten o'clock the next morning. But the milk was to be pasteurised before delivery to people's doorsteps and, unlike today, the temperature at which milk was stored didn't seem to matter.

George Ewart Evans also knew quite a bit about cows. Although best remembered as an oral historian, writing down the stories his friends and neighbours told him of how life had been when they were growing up in rural East Suffolk, one of his earliest childhood chores was to milk the family cow. His writing and life in rural Suffolk were informed by his own childhood experience of his family keeping a cow, as well as by the interviews he conducted.

Today most cows live on a farm, but a hundred years ago the Evans family were far from alone in keeping a cow in their backyard. This would probably have been a Welsh Black, a dual-purpose breed, good for both milking and beef. In the early years of the last century most cattle kept were of breeds local to the area. In Suffolk that was the Red Poll, which had evolved in the early nineteenth century as a cross between the earlier Suffolk Dun and its larger neighbour, the Norfolk Red. From the late nineteenth and early twentieth centuries, as farming became more intensive, higher-yielding breeds such as the

Friesian became more popular. Today those indigenous breeds are largely only kept by rare-breed enthusiasts, with the Jersey and Guernsey perhaps being exceptions.

Most of the milk produced by house cows was consumed by the family, either drunk raw or used to make butter and cheese. Consuming raw milk, something I did throughout my late teens, is not without its risks. Talking with Blaxhall historian Rodney West, I learned that Evans had contracted tuberculosis in the early 1950s while living in Blaxhall, from the milk delivered each morning to his doorstep by Fred Mayhew, the village milkman. The practice then was to bottle milk from a farm in the village, bypassing the pasteurisation that a commercial dairy would have employed before delivering milk to each customer's doorstep. Fortunately antibiotics quickly cleared up the disease.

A hundred years ago, a cow would typically yield around 4,000 litres of milk each year, so about half what they do today, but this was still more than enough to supply a large family with all they needed. Charles Hancy, who was born in 1899, described to Evans how his mother carried milk from their family cow round Bungay each morning. People did not have a fridge in those days, so would buy just what they needed that day, perhaps just half a pint. His mother charged just a penny a pint, so the money she made was hard earned. Charles and his sisters would milk eight cows before they went to school each morning, fetching them from the common outside Bungay where the family lived, then, after milking, taking them back out again to graze.

The Evans family cow was kept in a stable behind the shop his father owned in the Glamorgan mining village of Abercynon. In his autobiography Evans wrote about how much of a chore it was to look after the family cow, taking her out from the shop each morning to graze, then walking her back again to be milked in the afternoon. This had been his older brother

Gordon's job until he was called up. Evans was born in 1909, and his brother Gordon in 1901, so Gordon must have been called up in the closing months of the First World War. Evans would have been no older than nine when asked to look after the family cow. I was fourteen before I first milked a cow; Evans started far younger.

When Evans wrote his autobiography *The Strength of the Hills*, the stories of his childhood were dominated by memories of the family shop being forced to close in the mid-1920s – a casualty of the recession that saw unemployment soar to nearly 30 per cent in the Welsh mining villages.

However, he did describe something of the importance of the family cow elsewhere, in his novel *The Voices of the Children*, a story about a Welsh shopkeeper and his family. 'Outside were the stables, the big warehouse and the cart shed. There were two stalls in the stable: one for the horse and one for the cow. There were three regular customers for milk. We only had one cow, and we used most of the milk ourselves, selling only what was left over.'[1] The shop where Evans grew up is still trading today, and when I visited it in 2019 you could still see what would have once been the stable and cart shed to the rear.

In a subsequent passage in the novel, he continued, 'The first customer supplied with milk was Mrs Thomas, the teacher's wife. She was always tidily dressed, even in the morning. Her hair was in a net, and she spoke a little like Cardiff people.' This illustrated well how as a writer he drew on personal experience, even when writing fiction. He had been married to Florence for more than ten years when he wrote *The Voices of the Children*, and she, like Mrs Thomas, was married to a teacher (and was a teacher herself). When I knew her in the early 1960s, she also wore her hair in a net with a small bun, although, as a Londoner, she certainly did not speak with a Cardiff accent, Received Pronunciation being what was then expected of schoolteachers.

As you would expect, when a family kept a cow it almost inevitably became something of a pet. After all, it probably slept behind the house and kept you supplied with fresh milk for much of the year. The Evans's cow Daisy was, from what he wrote, quite strong-willed and had a tendency to escape from where she had been put to graze, to search out richer pastures. Evans described in his autobiography how he sometimes had to fetch her from the allotments behind the village where she was eating cabbages. This bovine wanderlust eventually led to her downfall. 'At last Daisy's wanderings came to a sad end,' Evans wrote. 'One day she broke down the fence and grazed for a long time on the lush growth down by the river. She got blown and had to be killed by the local butcher. I cannot remember my feelings at the time, but mother was very upset.'[2]

My father-in-law Michael kept a small herd of dairy cows at Church Farm, Blythburgh, in the 1950s, and not only were they sometimes treated as family pets, but many were also named after family members. This trend was started when a distant relative, Bert Latham, offered to buy Michael a cow shortly after he took on the tenancy in 1955. Then just seventeen, he was grateful for the gift, and happily complied with Bert's wish that the cow be named after his wife, May. After Michael married in 1958 he named one cow Winnie, after his mother-in-law, another after his wife Judy, and finally a new-born calf was named Belinda, after his daughter who is now my wife. The farm had fields on both sides of the main A12 road and even before she started school, Belinda would help her father take the cows across the road. Michael would stop the traffic, and Belinda would open the gate to release the cows. You could do that in the early 1960s, when there was much less traffic than there is today.

These stories only came to light when I found Michael's Register of Milk Records, an official-looking brown folder containing a page for each cow, upon which the milk recorder would carefully write down the milk yield measured on their monthly visit to the farm. Michael told me that at first, his milking parlour did not have electricity, so the vacuum pump for the milking machines was powered by an old Lister petrol engine and oil lamps provided lighting on dark winter mornings. Blackheath Estate had a similar Lister engine in their cowshed as a standby, which I can remember being used one day when I had accidentally taken down the overhead power line to the cowshed with the front loader on a Fordson Super Major tractor.

Being such a young farmer meant that Michael was spared National Service. Heads were scratched when his call-up papers were returned with the explanation that he was a dairy farmer and worked alone, and could not possibly leave his cows. Intriguingly, the previous tenant at Church Farm, Peter Wright, had won the Victoria Cross at Salerno in Italy in 1943; he took the farm tenancy on after being demobbed, before moving to a larger farm at Ashbocking at the invitation of Lord Tollemache, under whom he had served in the Coldstream Guards.

These memories prompted me to visit John Brown, one of Michael's contemporaries and a family friend since he and John were both Young Farmers. I took Michael with me and we drove to John's farm at Mells, near Halesworth, which straddles the River Blyth, so much of its 140 acres are water meadows. Looking up the valley from John's garden, the landscape reminded me of a Constable painting, with grazing cattle, a scattering of oak trees and the restored white-painted post mill at Holton visible in the distance.

John's father had been a tenant farmer at nearby Blyford, where they kept a milking herd of Friesian cows. Red House Farm, his present home, had been bought many years before and both were farmed together until his father's death. 'Had he lived longer, I could have taken over the tenancy,' John told me, but his father had died shortly before the 1986 Agricultural Holdings

Act made that possible. The milking parlour now stands deserted as John has replaced his dairy cows with a small herd of Limousin beef cattle. He took us in his twenty-six-year-old Land Rover to see them. I wondered why he was still actively farming at the age of eighty-three; he told me he believed that if he let the farm out, his daughters would incur a sizeable inheritance tax bill when he died. This, perhaps, is why so many old farmers continue to work long past retirement age, and why there are so few opportunities for young farmers to enter the industry today.

Dairy and beef farming require a cow to calve every year. When on heat, called 'bulling', other cows will jump on her back, simulating bovine sex. A combination of observation when bringing the cows in from the field to milk, and noticing the physical changes to the cow's rear end, which you tend to get close to when washing their udders before milking, would tell us that the cow was ready to meet the bull. This is far harder with beef cattle, which you might see from afar twice a day out in the fields, so John Brown runs a Hereford bull with his cows.

On dairy farms, more often than not, when a cow is on heat the AI (artificial insemination) man is called in. When I worked with cows in the 1970s, this service was provided by the Milk Marketing Board, but these days many farmers have been trained to do the job themselves. Others use a local contractor, often the cowman from a nearby farm with a large herd of its own. As a teenager, I found the process by which the cows became pregnant fascinating.

I can well remember the AI man who came to Park Farm, Friston. He would arrive from the Beccles Cattle Breeding Centre in a specially adapted Ford Anglia, the boot of which had been enlarged to take the large flask of liquid nitrogen, within which was stored the plastic straws of semen. These would have been collected from the bulls at the AI centre and frozen. Not only was AI convenient, but it also gave you a wider choice of bulls to choose from. This was important if the resulting calf was perhaps destined one day to be entered into cattle shows.

I cannot remember his name, but he was a cheerful red-haired chap of about forty five, with a skin condition that meant his nose was almost always red and flaky. This used to amuse me as I always associated his scaly nose with the fact that his job entailed placing it within inches of as many as fifty different cows' backsides each working day. The AI men were invariably referred to as 'bulls in bowler hats' both because of their job and the fact that they worked for the Milk Marketing Board.

When the AI man arrived, the cow would be waiting for him, tethered in the cowshed. He would put on a full-length rubber smock that fastened behind his back, remove the straw of semen from the flask in the boot of his car, fill a pail with hot water in the dairy, and then walk through to the cowshed and the waiting cow. Then, putting on a long disposable polythene glove that reached up to his armpit, he'd push his arm right up the cow's backside until past his elbow. With his other hand, he would then carefully insert a long thin metal tube into which he had already placed the straw of semen. As often as not he carried the insemination gun between his teeth until he was ready to insert it into the cow's vagina. A plunger allowed him to release the semen in exactly the right spot. This process sounds haphazard, but in fact had a high success rate. Most cows became pregnant, rather than come in heat again and need a second visit, known as a 'return'.

The deregulation of the milk market in 1994 and the introduction of quotas meant the end of the Milk Marketing Board, and today bull semen is sold to farmers by specialist breeders. Bull semen today cost around £25 a shot, with the farmer paying more for bulls with a more desirable genetic profile, for example higher milk yield or butterfat content. I was surprised to learn that, for the past fifteen years or so, it's been possible to buy sexed semen that is 90 per cent likely to produce a calf of the chosen sex. Bull calves, particularly Jerseys, are worth far less than heifer calves, so perhaps such early selection is a good thing.

The textbook gestation period for a cow is 283 days, but like babies, calves can arrive early or late. I used to marvel at the knack Fred the cowman had for knowing when a cow was about to calve. He would say that he'd noticed that she was 'bagging up nicely', which was his way of saying that her udder was filling out, ready to feed the soon-to-be-born calf, though of course most calves are taken away soon after birth. As the time neared, he would also say that she had started to 'drop', which meant that the ligaments on either side of her pelvis had relaxed, so her hips were more pronounced as the flesh sagged. Calves are relatively large when born, so the physical changes just before birth are usually obvious. Even a Jersey calf is likely to be born weighing around 28 kilos. That's as heavy as a healthy eight-year-old child.

But sometimes, particularly with an older cow, the changes before birth would happen quickly and she would have her calf out in the fields. I can remember how the whole herd would appear agitated and restless when this happened, with much sniffing of the new-born calf and mooing, as if congratulating the new mother on a successful delivery. More usually, the cow would be brought into the farm buildings before the birth, and shut on her own in a loose box.

There was something satisfying about helping a cow deliver her calf. Usually, they managed perfectly well on their own, but if a cow had been straining for some time with no result, Fred the cowman and I would lend a hand, feeling inside to see if a foreleg needed straightening and then using calving ropes to help ease the calf out.

Cows are herd animals, and so calves will stand up soon after being born, albeit on quite wobbly legs, and need little encouragement to find their mother's udder and start drinking. But this would not be for long, as we would want the cow to quickly rejoin the herd and be milked morning and afternoon. The calf would be taken away a couple of days after it had been born and placed with others in the calf pens, which at Park Farm were next door to the room where the cows' food was mixed, and that also opened into the cowshed where we milked. The mournful bellowing of the cow left us in no doubt that she was missing her calf.

But while taking the calf away soon after birth was a given in the 1970s, today attitudes are changing and some farmers are more compassionate. Not long ago I spent an enjoyable couple of hours with Stuart and Rebecca Mayhew, who have a small herd of Jerseys on their farm at Woodton, a few miles from Bungay. On his Twitter profile, Stuart describes himself as 'an ex-industrial farmer' who, as I can't help but think, is at the leading edge of what will become a growing trend. Throughout the last century, plant breeding, new fertilisers and agrochemicals have enabled farmers to increase their yields as each decade passed. At Old Hall Farm the Mayhews are reversing that trend, working in harmony with their land and adding value to their produce to compensate for the lower productivity their holistic approach delivers.

We met their cows standing in what until a few years ago was an intensively farmed wheat field. Now it is lush grass, with thistles and wild flowers because no agrochemicals are used. They don't use fertilisers either, instead allowing more grassland for each cow. I asked (with some trepidation) what happened to bull calves; in my day I could remember them having very short lives. I could see calves running with the cows, and learned that here they remain with their mother for up to a year, with the bull calves being reared for beef. I was surprised to learn after decades of being told that Jersey beef is too poor to market, that these days it is perfectly acceptable to the consumer.

I'd visited the Mayhews' farm a few years earlier, and what had been a small outbuilding containing a raw milk vending machine and a few locally grown vegetables was now a bustling farm shop and café. A young man working in the shop had gone to fetch Stuart and Rebecca when I arrived, but had first talked enthusiastically about the farm and his job there. There are probably as many people working on this 500-acre farm today as there were when George Ewart Evans interviewed men who'd kept cows in nearby Bungay a century ago.

Because the cows keep their calves, they are only milked once a day, but there is still enough milk to supply customers visiting the shop and also the growing number who buy raw milk regularly by mail order. It is far more profitable to sell your own produce than to do so through a wholesaler. The farm shop stocks meat from the farm, too, and a wide variety of home-baked pies and flans. As concern about climate change grows, it was encouraging to see this contemporary approach to farming, creating jobs, adding value, but not forcing the highest yield from every acre of land by applying chemicals.

When I worked on farms, female calves would be reared to replace old cows that had left the herd. Bull calves from most herds would go to market, to be reared by other farmers for beef, but the Jersey calves at Park Farm would be sold to the local hunt to be shot and butchered as food for the hounds. The price you'd get for them at the market barely covered the cost of taking them there, and we were told we could not afford to be sentimental; how different things are today.

The calves we reared would be fed powdered milk substitute, which ironically consists mainly of dried whey and skimmed milk, by-products of cream and butter production. One of my weekend jobs was to feed the calves, mixing the powdered milk with hot water in a pail, then pushing the calf's head into the bucket, letting it suck on one or two of my fingers submerged in the milk. A calf would naturally stretch up to its mother's udder, so lowering their head to drink is counter-instinctive. Pretty soon, though, the calf would become used to bucket-feeding and would drink unaided.

Cows produce milk for around 300 days after a calf is born, and any delay to the conception of their next calf stretched the dry period from the usual sixty days to ninety or more. Dairy cows are both pregnant and producing milk for most of their lives, which is why they never grow old. Heifers (young cows) are first served at the age of eighteen months, and after four or five lactations, both yield and fertility begin to fall. Although a cow has the potential to live for twenty years or more, few last more than six or seven years before going for slaughter. Jersey cows culled from the herd would end up as pet food, which means they were worth far less than a Friesian cow, which would end up in meat pies and other processed foods. Beef breeds, such as the Limousin or Hereford, produce better-quality meat such as steak, and joints for roasting, but little milk.

Back in Evans's day, there was far less pressure on profits, and I suspect that had Daisy, the family cow, not 'blown' and had to be put down by the village butcher, she would have lived a long and happy life in the stable alongside the horse, behind the shop in Abercynon. Even the dairy herds on the farms in and around Blaxhall would have been managed less intensively in the 1950s, when Evans lived there, than they would today. I wonder what he would make of the Mayhews' farm today, which is little more than two miles from Brooke, where he and Florence lived after she retired from teaching.

Until the early twentieth century, little had changed for centuries on farms and in rural villages. This point is well illustrated by some of the customs Evans observed, which will have been passed down through many generations. In *The Pattern Under the Plough*, the old farm workers he interviewed told him of one particular practice that followed the birth of a calf: 'In clearing up after the birth, the stockman took the placenta and threw it over a white-thorn, usually on a remote part of the farm, where it remained until it rotted away.' This Evans described as 'pure contagious magic', because it was thought that doing this would transfer the vigour of the bush onto the cow and newly born calf. As he wrote, 'the thorn-bush was the "quickset", which as its name implies, was abundantly alive. This would ensure that the animal would still remain quick and fertile and would breed again the next season.'[3]

As with so many traditions, each new generation of stockman would learn from his predecessors, and, even if they doubted the value of the ancient customs, would have been reluctant to stop following them, just in case they made a difference. Those old stockmen must have been quick though, or perhaps each had fewer cows to care for, because when I've watched cows calving they invariably eat the afterbirth pretty soon after the calf stands up. This is an inherited bovine instinctive action that removes evidence of a recent birth in case predators spot it and hunt the area for the calf, which for the first few days will be unable to run away.

Another custom that Evans wrote about, which was probably centuries old, was that if a cow slinked (aborted), her calf would be buried in a gateway as a matter of course, as a precaution against contagious abortion. The old stockmen believed that as cattle walked through the gate, above the calf's body, they would somehow be protected from their own as yet unborn calves suffering the same fate. Dairy cows were important to the rural economy, and before vets, folklore was all people could rely on.

In wintertime, when there was no grass to graze, cows were fed a mixture of fodder beet and bran. The fodder beet was grown over the spring and summer, then stored in a clamp (an outdoor heap covered with straw) to inhibit sprouting and prevent rotting. Bran would be bought from the local mill as a waste produce from flour making. Most villages had a wind-powered flour mill that would buy locally grown wheat and sell flour. In Blaxhall and the surrounding villages, they will almost certainly also have fed their cattle with malt culms and barley screenings from nearby Snape Maltings, then a thriving enterprise, and today a concert hall. Nothing was wasted in those days.

The fodder beet had to be chopped up before it could be fed to the cows. Charles Hancy described the process to Evans, telling him how he would put the beet into a hopper at the top of the grinder, then laboriously turn the handle to shred them into a pail beneath. Then the shredded beet would be mixed with chaff and fed to the cow. He explained how, as well as milking the cows before going to school, he would return at lunchtime to grind fodder beet and add bran ready for the afternoon milking when he had finished the afternoon's lessons.

He was ten years old; his friends chose not to sit next to him at school because he smelled of cows, having helped tend his family cow before setting off to school. He can't have been alone though, as he said there were ten Bungay families living on Nethergate Street, where he lived, that kept a cow, and ten more on Broad Street. Bungay sits next to the River Waveney, along which there are acres of water meadows, too wet for growing anything but grass. These, as well as the common, would have been where everyone grazed their cattle.

At a time when people had little money to buy food, families living in rural villages such as Blaxhall were largely self-sufficient, growing vegetables, keeping a pig, chickens and perhaps a cow too, as well as gleaning the harvest fields in summer for wheat to grind to make bread. Gleaning dates back to biblical times, when grain was deliberately left on the ground for poor people to pick up. Milk came either from your own cow, was bought from a local farmer, or perhaps the village shop, and was an affordable staple in the country diet. I was reminded of my visit to Old Hall Farm where, once more, local people were buying locally produced milk, meat and vegetables. The Covid pandemic made it less attractive to visit a large supermarket, where you might encounter more than a hundred other shoppers. Many rediscovered the more relaxed experience of shopping closer to home.

Today, UK milk consumption is falling, with government statistics showing that on average we each consume seventy litres annually, while in 1970 we drank twice as much. This is partly a consequence of falling consumption of so-called 'host foods' such as breakfast cereals, and partly because of increasing health concerns, in particular lactose intolerance. As demand for fresh milk continues to fall, so too does the number of dairy cows. *Farmers Weekly* reported that in 2020 there were 50,700 fewer dairy cows than in the previous year. It's no longer enough to strive for greater efficiency, pushing up milk yields and driving down costs. The smart dairy farmer today adds value to his milk rather than simply selling it in bulk to a wholesaler.

Perhaps one of the best examples of this in Suffolk is Fen Farm at Flixton, on the outskirts of Bungay, a town where, as George Ewart Evans wrote, many people kept a few cows. I visited another third-generation farmer, Jonny Crickmore, and we walked up the steep bank behind his farm, from where we could see the cows grazing on the meadows that run alongside the River Waveney. With 300 cows and few trees, the view was a little less bucolic than the pastoral idyll that was John Brown's small farm some ten miles further into Suffolk, but attractive all the same.

Jonny did not go to college or university, because, he told me, he just wanted to be a farmer. But only just into his forties, he was one of the most innovative farmers I've ever met. Recognising the dangers of selling all his milk to a single wholesale customer, and with riverside meadows that can only grow grass, he imported some Montbéliarde cows from France a few years ago and started making a Brie de Meaux cheese called

Baron Bigod from unpasteurised milk on his farm. It's long been one of my favourite cheeses and sells at Neal's Yard and Harrods in London, so I was particularly keen to visit the farm where it was made.

The afternoon milking started while I was on the farm, and it was impressive to see the cows in the parlour, twenty-eight at a time, with milking-machine clusters that automatically withdraw when each cow has been milked. As the cows leave the parlour, they pass a sensor that can read the transponder each cow wears. If a cow needs to be kept back, perhaps to see the vet, an automatic gate diverts it into a pen rather than being released back into the field with the others. This saves a lot of time and effort.

As with the Mayhews' farm, Fen Farm now employs many more people than it would if simply producing milk. Baron Bigod cheese has become popular and many tonnes are sold each year both in the UK and, increasingly, around the world. Characters in BBC Radio's farming soap *The Archers* even bought some Montbéliarde to make a raw blue cheese, and to this end a BBC researcher contacted Jonny to find out exactly what making the cheese involves.

Those George Ewart Evans interviewed, who told him about the old ways they remembered, were a little nostalgic for the practices they saw slipping into history. The years that followed saw milk and cereal yields rise, pushed ever higher by fertilisers and sprays, with bigger machines needing few people to farm even a thousand acres. Few in the future will look back with nostalgia at the way things were over the past few decades, because the future that is starting to emerge is far brighter, much kinder and, once more, the number employed in the rural economy is increasing.

Wheat

FORDSON TRACTORS HAVE FEATURED in my life for more than fifty years. Fittingly, I started out at the age of fourteen driving the smallest in the range, a thirty-two-horsepower Fordson Dexta, when I worked with the cows at Blackheath Estate. It was tiny compared with today's tractors, the largest of which can be twenty times more powerful.

A regular Saturday seasonal job was to hitch a large four-wheeled trailer to the little Dexta and drive three miles to Billeaford Hall, Knodishall, where I would load it with wheat straw from an outlying Dutch barn, then return to Park Farm at Friston. When loaded, the trailer was really too heavy for the tractor, and to brake too sharply when going downhill could easily have resulted in the loaded trailer pushing the tractor over. But the Dexta was not a fast tractor, and so I never needed to stop quickly.

The wheat straw was used as bedding for the cattle and would be spread in the yards where they spent the winter months. This was called littering, and as the months passed, the straw and cow-muck mixture would build up to be a good three feet deep. In spring, when the cows were turned out to grass, another regular task was to muck out the yards. A larger Fordson tractor, a forty-two-horsepower Major with a front loader, would clear most of it away, but there were always parts the tractor could not reach, so these had to be cleared with a fork and barrowed to where the tractor and loader could collect it.

Wheat straw has always been used for bedding-down animals. Barley and oat straw are softer and so more easily digested, providing necessary fibre in the bovine diet. Both are fed to cattle and horses, while wheat straw is only useful for bedding, but even then, it is less absorbent than barley straw, so more is needed.

Years after my time at Blackheath Estate, in 1977 when I was on my sandwich year as an agricultural college student, I worked for Henry Greenfield at Crown Farm, Leiston. As an arable farmer he had the worrying habit of burning fields of wheat straw without first cultivating the headlands round the field to prevent the fire from spreading.

He'd drop a lit match in one of the swathes of straw left by the combine harvester, and let the wind carry the flames across the field and often into the gorse and bracken that lay beyond. This accidental burning of the surrounding heathland was, he said, a good way of controlling the rabbit population, cheaper than putting cyanide powder into their burrows, which was another of my summertime jobs that year; I used an old dessert spoon nailed to a broom handle for this, and a spade to block the hole.

Burning straw in the field was considered a good way to control black-grass, a weed resistant to many herbicides, but the practice was banned in 1993 because of growing environmental concerns. Today many combine harvesters chop the straw, but an increasing amount is baled and sold as fuel for biomass-burning power stations. This is a relatively new industrial market for an agricultural by-product.

Wheat is the most widely grown cereal crop in the Western world. In the UK alone, around 15 million tonnes are grown each year. This is more than twice the combined tonnage of barley and oats. Wheat is the grain used to make flour, which in turn is baked to make bread. According to the UK flour millers' trade body, 99.8 per cent of all British households regularly buy bread. 'Man cannot live by bread alone', but its popularity suggests that he can't live without it either!

Wheat has been grown in Europe for more than 10,000 years, originating when early farmers repeatedly saved and replanted the largest grains from harvested wild grasses. Over the centuries, these evolved into the wheat we know today. It is no wonder that those George Ewart Evans interviewed for his books spoke about wheat. Priscilla Savage, who lived next door to Evans in Blaxhall, baked bread twice a week. 'Prissy's family of 10 had two bakes a week, using altogether five stones of flour,'[4] wrote Evans. 'The flour was often mixed with the whey left after making butter or cream cheese at the farm.'[5]

What Priscilla Savage's stories of life before the First World War illustrate well is the way that life in rural Suffolk was largely self-sufficient. The flour she used would have come from the village mill; the yeast was saved from the last brewing of beer; Blaxhall was surrounded by heathland so gorse was plentiful and provided a ready source of fuel. Like all cottages at the time, hers had a bread oven that was heated by lighting a fire inside it. No thermometer was used to check the temperature; instead a handful of flour was thrown lightly against the side of the oven, and if it burned up with a blaze of sparks, the housewife knew the oven was hot enough for baking. The fire was then raked out and the tins of dough placed inside and baked with the heat retained by the bricks around the oven.

Evans observed that this method of baking bread had hardly changed since biblical times, noting from Matthew's gospel: 'the grass of the field, which today is alive and tomorrow is thrown into the oven' – a technique little different from the one Evans saw his neighbours use. He also noted that the older people he interviewed were nostalgic for bread as it used to be – stone-ground and often contaminated with flecks of charcoal from the oven, which they claimed gave it extra flavour. This was very different from the factory-baked, shop-bought bread they were then eating.

Some Suffolk bakeries, though, have changed little over the years; Stradbroke's bakery has been making breads continuously for more than 200 years. In the early days, the baker used to allow villagers to use the still-warm oven to cook pastries and joints on a Sunday, although this practice ceased in 1860 when William Rutter took over. A staunch Baptist, and deacon of Stradbroke's Baptist church, he could not condone such activity on what he saw as the Lord's Day.

The bakery has survived until today because it now supplies a number of other village shops, rather than being totally reliant on sales from its own shop. That said, it always seems busy when I drive past on my way to the coast. While Stradbroke has a population of some 1,500, it's all too easy to drive to nearby Diss with its two large supermarkets.

An even smaller bakery was the one I visited at Wakelyns, a small organic farm at the end of a very long single-track lane, near Fressingfield. It was bought in 1992 by plant pathologist Martin Wolfe and his wife Ann. He practised agroforestry, which involved dividing the fields up into strips, which he called alleys, with each separated by trees, some of which are coppiced, and some fruit. Some of the alleys are cropped with YQ wheat, which was bred by making 190 crosses of twenty different varieties selected for their yield (the Y) and quality (the Q). Seed is kept for the following season, and the wheat has adapted to suit its habitat.

When harvested, the wheat is stored on the farm, ground and used to bake sourdough bread at the on-farm bakery. While there I enjoyed a slice spread with butter from Fen Farm, Flixton, ten miles away, and jam made from plums grown at Wakelyns. This was hyperlocal food production, with the bakery supplying a number of local shops, including the community-owned shop at Metfield I was to visit later. Wakelyns run a programme of courses, as well as providing accommodation for people wanting to spend a few days getting closer to nature.

As well as shopping in the village, many of those Evans interviewed also remembered going gleaning. It was usual for the miller to keep some of the flour in lieu of payment.

In common with so many aspects of late-Victorian village life, in Blaxhall the church played a role in the gleaning process as 'the gleaners – usually women and children – were allowed only to operate within the bounds of the parish'. Evans described how the field being cleared of sheaves of wheat was the signal that the gleaners could start work. The church bell was rung at eight in the morning, then again at seven in the evening, to mark the start and end of the gleaning day.

In those days, the bread that you ate would have been baked locally, from wheat grown and ground into flour all within a couple of miles of your kitchen table, just as I'd sampled at Wakelyns. This contrasts starkly with today, when the wheat might have been grown in North America, milled at Tilbury and baked at Enfield before being delivered to your nearest supermarket. This change from local to global was starkly illustrated when I met Tim Hirst, who had been involved in the grain trade for more than fifty years.

The grey Ferguson 20 tractor standing in his garden hinted

at Tim's undiminished passion for farming. He's clearly a sociable man, too, and I suspect guests staying at his Airbnb, a converted farm building next to his house, find him a convivial host. We sat in his garden and over coffee he told me how, since entering the trade in the mid-1970s, he has seen the company he's worked for taken over several times; today, he said, there are really just four main players in the UK grain trade.

Comparing career stories, I realised that Tim and I had come close to meeting in the 1980s when he had been seed and fertiliser manager for BDR, a merchant business based at Bourne in Lincolnshire. The company had been agents for Hadfields, a direct-selling brand of Fisons Fertiliser, and I had been for a short while the Hadfield's company rep in that area. On realising this Tim disappeared for a few minutes and returned with a mounted plaque, awarded to him in 1990 as Hadfield's top agent of that year. This coincidence reminded us both that the farming world was a small one. But for a promotion in 1984 that took me to North Yorkshire, I would probably have been the man who presented Tim with the award.

We talked on, remembering how there were once many reps calling on farms. With no mobile phones, business was transacted in the farmyard, at the farmhouse kitchen table, or sometimes in the middle of the field. I'm sure that Tim had, like me, been adept at driving his company car up bumpy tracks and across fields. We discovered that we had both also spent time as cereal crop inspectors, having been trained in identifying the variety of cereal from the physical characteristics of the ear of corn. An exam was involved, at the National Institute of Agricultural Botany in Cambridge, and led to summer days walking fields of seed wheat and barley, checking that the crop was pure and free of wild oats.

Today, although now in his early seventies, Tim still looks after sixty farming customers. A full-time rep would have had 200, far more than would have been the case in the 1970s. Reps now rarely visit the farm. Today business is mostly transacted by phone, with grain prices dictated by global events. Tim starts each day by checking the world markets and sends his customers an email bulletin before lunch. Despite this move to digital communication, he said the trade remained a people business, with farmers loyal to their trading partner, often for generations.

While this has been the case for many years, as more farms are bought by successful city entrepreneurs and contract farmed on their behalf, these links with tradition are gone and grain deals often won or lost for a few pence a tonne. Everything today is bigger, faster and, dare I suggest, more business-like. Reflecting on this later with my father-in-law Michael, he said that farming had turned full circle. Once, he said, the landed gentry owned the land and tenants, like he had been, farmed it. Today it is new rather than old money that owns the land, and contract farming a new alternative to tenancy. One contract farmer might work with a number of landowners, using massive tractors and employing few people.

George Ewart Evans could see that life in rural England was changing, and this was what prompted him to write down the stories his neighbours told. I suspect he realised that change was accelerating, and this gave some urgency to his work. From Blaxhall the Evanses moved to Needham Market, and a few years later to the estate or 'closed' village of Helmingham, which had been owned by the Tollemache family for more than 500 years.

Farm workers here had for centuries grown wheat in the gardens behind their cottages. Each was a half-acre in size, divided in half by a path down the centre. They worked their gardens by what Evans described as a 'two-year rotation', growing wheat on one side and vegetables on the other, changing sides each year. The wheat would be sown and harvested by hand, and then threshed with a flail on a tarpaulin laid out in the cottage garden.

At the bottom of each garden was a pigsty that provided both meat for the kitchen and manure for the garden. The pig would have eaten any scraps from the table, as well as fodder beet that were also grown in the garden. The garden, large by today's standards, was created to encourage each cottager to be as self-supporting as he could. Few today will remember keeping a pig or growing wheat in their back garden.

The increasing mechanisation of farming through the early years of the twentieth century even extended to the gardens behind those farm workers' cottages. Joe Thompson told Evans how in the late 1920s he had bought a two-coulter (two-row) seed drill that he restored and then hired out to his neighbours. Edith Fox (born 1891) remembered her father using it, telling Evans how a man always pulled the drill, but never an animal, even though there was room for a donkey or pony to have been used.

But although faster than hand sowing, using the drill did not save that much labour. Joe Thompson described how in 'one short afternoon, four of us drilled an acre of wheat . . . There were two pulling, one in the shafts of the drill, another hitched up like a trace-hoss, the third walking behind the drill to see it was running right and the fourth raking over.'[6] On the farm, a Smyth drill would have been used to drill wheat. It would be pulled by one or two horses and had twelve coulters across its eight-foot width. The hand drill, with just two coulters, was big enough to take real effort to use, but small enough to comfortably drill wheat in the cottage gardens. Today, many smallholders will use a lightweight single-row seed drill that can be pushed by one man and can be adjusted for a wide variety of seed sizes and rates.

Just as Edith Fox said she'd never seen an animal used to pull the small seed drill, she would not ever have imagined a woman pulling it either. Perhaps it was the fact that outdoor work was physically demanding that meant that it was always the men that did it, or perhaps it was simply that until the twentieth century, gender roles were never challenged. (Even in Neolithic times, it seems that when they died, men were buried with arrow heads and women with ceramic bowls.)

Priscilla Savage's granddaughter Daphne Gant had told how she and other Blaxhall women had put in long hours picking and sorting potatoes, and other manual jobs on the farms. Her husband John's wages had paid the rent and put food on the table, but it was Daphne's hard work that allowed the couple to buy a fridge and television set. Was it really less arduous to spend the day picking potatoes than to walk behind a horse-drawn plough?

Reflecting on my conversation with Daphne made me think about how my mother had stopped work when I was born and for the rest of her life had been what she happily called a house-wife. She played a supporting role to my father as his career at the bank progressed, but did not earn a penny herself. Did that mean that we were middle class because we were able to live comfortably on just one income?

My mother-in-law and her twin sister never really worked either, both marrying local farmers, both raising two daughters and both running countless errands for their husbands. My wife had a responsible job for twenty-five years, once the children started school, and our daughter has a successful career as an epidemiologist. Nothing would horrify her more than the thought of giving up her work to raise children and bake cakes. Attitudes have changed so much in just three generations.

George Ewart Evans had a writer's curiosity and in the early 1960s, when living at Needham Market, he was intrigued by the history of the house they lived in, which was built in the sixteenth century. 'One of the rooms,' he wrote, 'is beautifully panelled but a former owner told me that, sometime before, they had laboriously to scrape off the whitewash, with which the panels had been thickly covered.'[7] The house had been a bakery, and the panelled room was where the flour had been sieved. Public health regulations meant that the walls of bakeries had to be whitewashed every three months and so the panelling in the room had been whitewashed many times.

Wheat yields have grown substantially since the 1890s, when those Evans interviewed grew up. In 1885 a yield of one tonne per acre was considered good, and by the 1940s, this had only increased by around 10 per cent. Then in the mid-1960s, ammonium nitrate fertilisers were developed, and over the next twenty years yields grew to around three tonnes per acre. Today four tonnes of wheat per acre is not unusual. In the late 1970s, ICI Fertilisers launched a 'Ten Tonne Club', providing recognition for farmers achieving yields of ten tonnes of wheat per hectare (which is four tonnes per acre). However, in the 1979/80 season, only sixteen farmers qualified for membership.

Years before fertilisers were invented, farmers knew that they needed to rest the land between wheat crops. 'The Anglo-Saxon and later the medieval farmer fallowed or rested a third of their arable land every year,'[8] wrote Evans. 'Cattle and sheep fed on the weeds and grasses that grew on this fallow land,'[9] which was also ploughed three times to prevent the weeds from going to seed.

'There is no question of the merit of fallowing, when compared with bad courses of crops,'[10] wrote Evans when reflecting on his research into the way early nineteenth-century farmers managed the land. Weed control was achieved by sowing 'two successive hoeing crops,'[11] usually beans and cabbages, which, being planted in wider rows than wheat, were easier to hoe. This practice would have been familiar to the medieval farmer too.

In the early 1700s wealthy Norfolk landowner Charles Townshend formalised this practice on his Raynham Hall Estate, introducing a four-course rotation for which he is remembered today. Wheat was followed by turnips, then barley, and in the fourth year grass and clover. Because most farms had livestock, the turnips provided valuable winter feed. Grazing the grass had the added benefit of adding manure; sheep were described as providing the golden hoof, making the subsequent cereal crop more abundant. Muck from the yards where cattle overwintered was also spread on the land that was going to be sown with wheat. His innovative new way of farming earned him the nickname 'Turnip Townshend'. Today the nearby Holkham Estate is experimenting with adding grass and clover back into its rotation, describing this as rejuvenating the soil, improving its structure and building soil organic matter back-up.

Medieval landowners, the manorial lords, also took control of the milling of wheat grown by their tenants. Since ancient times, wheat had been ground at home using querns – pairs of flat, rough, round stones – between which the grains were placed. By rotating the top stone, called the turnstone, the wheat was ground into coarse flour. Water mills were first used in the tenth century, to turn larger millstones and so produce flour more efficiently. 'By the time of the Domesday Survey (in 1086) there were between five and six thousand water mills in England.'[12] The population of England was then below 2 million.

These watermills were not cheap to build, and millers had to be employed to run them. To make sure of a return on their investment, the manorial lords claimed sole rights to grind corn in their jurisdiction. Querns were outlawed and, if discovered, smashed. People now had to pay to have their wheat ground; the miller would keep some of the flour he ground in lieu of payment, just as the miller in Blaxhall would have done centuries later. This he would then sell to those who did not grow wheat of their own.

Later, post mills, powered by the wind, became popular, perhaps because in summer the rivers ran low, but in the UK there is usually enough wind to turn the sails on the windmill. Across most of the UK, average wind speeds over a year are around ten miles per hour. That is enough for a windmill to operate and today, enough to power a wind turbine generator. In areas with less wind, you just need larger sails on your mill, or larger blades on your turbine. For many years I had two five-kilowatt wind turbines behind my home, and they provided around half the electricity we used.

The sails of a post mill were originally just that – canvas sails lashed to four wooden frames. The earliest mills were built without the brick tower you most commonly see today. This was to make it easy for the miller to attach the canvas, as Evans explained: 'To enable the miller to do this, the arms or stocks swept to within a couple of feet of the ground.'[13] As you might imagine, this was not always safe, as in a strong wind the sails would turn quite quickly. Mounting the sails further from the ground required a change of design. Today the mill 'sails' are wooden vanes whose angle can be adjusted from within the mill. These vanes could be opened or closed according to the strength of the wind, rather like the slats of a Venetian blind.

The first mill I can remember visiting was at Saxtead Green, near Framlingham. Now in the care of English Heritage, it has long been open to the public. I suspect we visited from Needham Market, on one of the family outings my father organised when he first owned a car. The present mill was built in the late eighteenth century, although there had been a mill on the same site centuries earlier. Painted white, it is hard to miss as you drive along the road from Stowmarket to the coast. It was on the route we always seemed to take when driving down from Yorkshire to visit my wife's parents in the 1980s.

Saxtead Mill was privately owned by the same family for its last century of operation. In 1947, an oil-burning engine was used to power the mill, rather than relying on wind power, and in 1959 it closed altogether. Some mills, for example at

Pakenham near Bury St Edmunds, and Letheringsett near Holt in Norfolk, have enjoyed something of a renaissance in recent years. There are now just twenty-four working water mills left in Britain. Letheringsett Mill was built in 1800 and is powered by the River Glaven, producing a respectable four or five tonnes of flour each week. Visitors can tour the building and see how all flour was once produced. You can buy artisan bread flours, both from the mill's shop and online. This is an interesting juxtaposition of a very old way of milling wheat with e-commerce, a very modern way of doing business.

Most flour today is produced by large mills that are in corporate ownership, although one of my classmates at agricultural college in the mid-1970s was a scion of the Marriage family, who have been millers since 1824 and have managed to remain a family firm to this day. The company has an annual turnover of more than £50 million and supplies commercial bakeries across Essex and beyond. They also make animal feed and farm on the outskirts of Chelmsford, a few miles from the mill. According to the National Association of British and Irish Flour Millers, just thirty-two companies between them produce the 5 million tonnes of flour consumed in the UK each year. They also say that the UK is almost self-sufficient, growing 84 per cent of the wheat it needs.

Something that the people Evans interviewed in the 1950s could never have imagined is the extent to which wheat is now used industrially. Today 1.5 million tonnes of wheat are used annually to make starch and bioethanol. Starch is used in processed foods, but almost half of the starch produced is used in paper-making, where it makes the surface stronger and smoother, so easier to print or write on. Demand for paper, particularly newsprint, is falling, and each year the amount recycled increases. Demand for bioethanol is, however, increasing.

Wheat for bioethanol is ground as if to make flour, then mixed with water and heated. Enzymes are added to convert the starch into sugar, then this is fermented to turn the sugars into alcohol, which is finally distilled to create bioethanol. The residue from this process is rich in protein and is pelleted and sold as animal feed. Unleaded petrol in the UK currently contains 5 per cent bioethanol, with the government proposing to double this to 10 per cent.

Bioethanol is also manufactured from sugar beet, with the beet factory at Wissington in Norfolk producing more than 60,000 tonnes annually, as well as 400,000 tonnes of sugar, all from locally grown sugar beet. Sugar consumption is falling, in response to public demand and Government pressure to reduce the amount that is added to manufactured foods and drinks. Evans's work charted the change from horse to steam, and then tractor. Perhaps future historians will write about the early years of the twenty-first century as the time when agriculture became a significant producer of industrial raw materials, rather than just growing food.

Just as farmers are now growing wheat and straw for industrial processes, so too is growing food now taking place away from the farm. British Sugar uses waste heat from the Wissington sugar beet factory to grow tomatoes and medicinal cannabis in glasshouses that cover forty-four acres. Sugar beet processing is seasonal, taking place between September and March each year, enabling a winter crop to be grown as well as a summer crop. Near Norwich, a similar-sized area of glass has been erected on the Colman family estate and heated by the city's nearby sewage works. One hundred and twenty years ago, you could make a living from a forty-four-acre farm and the average holding was less than eighty acres.

Wheat has always been a traded commodity. Farms in rural communities such as Blaxhall will have grown wheat that was ground at the village mill and used by villagers to bake bread. But more will always have been grown than is consumed locally, so surplus production will have been sold on the open market.

Wheat is also easy to store and does not deteriorate if kept dry. As towns and cities grew, so too did the demand for wheat and flour. Much of the wheat traded from Blaxhall will have found its way to Ipswich, where the Corn Exchange, built in 1882, replaced the open-air grain market that had been held on Cornhill for centuries. Here buyers and sellers of wheat would meet, examine samples of grain and strike deals. The repeal of the Corn Laws in 1846 opened up the market, which previously had been protected by tariffs from cheap imports.

In my youth, farmers sold their corn to agricultural merchants, who in turn sold it on to flour millers, maltsters and animal feed mills. Tim Hirst talked of how he would collect a sample from the farm and take it to the mill himself, where the price offered was based on a visual inspection, rather than, as is the case now, a laboratory test. He would then return to the farmer and offer to buy so many tonnes, building in his margin, and arranging transport from farm to mill.

Mixed farms produced pigs, beef cattle and perhaps milk all year round, which provided more regular income. Cereals, potatoes and sugar beet are all harvested from late summer into the autumn, so it was not unusual to see the merchant's lorry loaded direct from the combine in the field. On my father-in-law Michael's farm, the merchant would sometimes leave a lorry to be filled on the field, returning to collect it the next day. This gave me the opportunity in my late teens to drive an eight-wheeled lorry up the road back to the farmyard. Of course, I did not have a heavy-goods driving licence . . .

Over time, a futures market for grain evolved, so that farmers could sell their crop before it was harvested. This worked well when the yield and quality of the grain, when it was cut, was equal to or better than what had been sold. Penalties were paid if the quality fell short. Similarly, many farmers would store grain for many months after harvest, selling it later when scarcity meant it would fetch a higher price.

Wheat is now a globally traded commodity. It is the most commonly grown cereal crop in the world, with more than 700 million tonnes grown each year. While the Corn Laws were passed in the early years of the nineteenth century, to protect UK farmers from cheap continental imports, by the end of the century North America was exporting wheat to the UK. The cold winters and long, dry summers in the American Midwest are ideal for growing hard-milling wheats, which contain more gluten and so are better suited to bread- and pasta-making than the softer wheats more commonly grown in Europe.

When I left college, I joined fertiliser company Fisons, so worked alongside the agricultural merchants that stocked and sold our products. Their sales teams sold fertiliser, seed and agrochemicals to farmers and my job was to advise both the sales reps and farmers on agronomy. In the late 1970s, this was reckoned to be a good way of maintaining my employer's market share. Farmers also sold their grain to the merchant from whom they bought their supplies. As Tim Hirst had reminded me, relationships and loyalty were more important than price; although haggling was inevitable, the favoured rep would always be given the opportunity to match the best deal the farmer had been offered.

The merchant reps rarely changed jobs and their relationship with the farmer was close. I can remember being surprised the first time I saw one go into the farm kitchen without knocking, put on the kettle and make a pot of tea. They were treated like family, and their intimate relationship with the farmers buying and selling created a sense of mutual dependency.

For the farmer with pigs or cattle, the relationship with the merchant rep was even closer, as they would both buy their grain and sell them animal feed. When I was selling to farmers, there were still local feed mills operated by agricultural merchants. The feed that a farmer bought might well have been formulated using grain from his farm, with protein and minerals added to create the balanced diet the livestock needed. Fishmeal was a common source of protein, giving the feed a distinctive odour. In the early twentieth century, when those Evans interviewed were young men, fishmeal was more likely to be used as fertiliser.

Looking back on my years in the agricultural supply business, I can see that I was witnessing the very tail end of the local trading that had been in existence for centuries. Today most of the agricultural merchants and local feed mills have disappeared. There are also fewer farmers, with, for example, most of the land around Blaxhall now farmed by a Sussex-based farming group that grow vegetables and herbs, as well as more conventional crops, on more than 6,500 acres.

Have all these changes made rural life better? I doubt it. Farming is now capital-intensive, usually on a large scale, but still heavily reliant on the weather. This makes it a risky and often lonely occupation as it no longer takes a team of men to run a farm. We are social animals, happiest when members of a small community. Our way of life may have changed, but we have not.

Wool

Thirty years ago we took a family holiday in Denmark. The children loved visiting Legoland, where I can remember our son caused havoc at the Lego driving school by driving on the left-hand side of the road while all the other children drove on the right. But my lasting memory of that holiday was our visit to the museum at Silkeborg, some forty miles from Legoland.

There, in a dimly lit room, I saw Tollund Man, who had died more than 2,000 years ago and been discovered in 1950 preserved in a peat bog. He looked as if he was asleep; not at all like the victims of the eruption of Vesuvius we'd seen on an earlier visit to Pompeii, their obvious distress preserved for ever by the volcanic dust that had engulfed the town. However, Tollund Man had not died peacefully as he had been hanged, either in punishment or as a human sacrifice; he still had the noose round his neck.

Perhaps the most surprising thing about Tollund Man was that despite having lived so long ago, he was clean-shaven and his hair was neatly trimmed. He was also wearing a wool-lined leather cap. Clearly pre-Roman Denmark was far from uncivilised, which perhaps was almost as startling a realisation as coming face to face with a man who had lived so long ago.

People have been keeping sheep since well before the time of Tollund Man. Sheep are small enough to manhandle, relatively docile and versatile, providing wool, milk, meat and leather. Garments have always been made from sheepskin, from ancient times right up to the present day. Living as we do in northern Europe, wool is particularly important as it can keep us warm.

George Ewart Evans wrote a lot about sheep, or rather about the men who worked with them. His next-door neighbour in Blaxhall was retired shepherd Robert Savage, who took great pride in telling Evans about his life and work. Savage had worked with sheep from the age of sixteen, when he moved from being what was known as a back-house boy, working in the kitchen, to helping the shepherd full time. He enjoyed looking after the sheep, and, when the shepherd was sick, took charge of the flock, for which he was paid a man's wage.

I wanted to gain an insight into shepherding and so arranged to meet Robert Savage's grandson Ivan, who had from the age of fifteen become the seventh generation of his family to work with sheep. Appropriately, Benjamin Britten was BBC Radio 3's Composer of the Week on the day I drove down the A140 to meet him and his daughter Mandy at the new Suffolk Archives building called The Hold. Presenter Donald Macleod played extracts from *Peter Grimes* and *The Young Person's Guide to the Orchestra*. He also talked about Aldeburgh beach and Snape Maltings, and by the time I arrived at The Hold, I was feeling quite nostalgic.

Ivan Savage is old enough to remember his grandparents Robert and Priscilla Savage; he had been ten years old when his grandfather Robert died, an event Evans had touchingly written about in his autobiography *The Strength of the Hills*: 'I was fond of Robert Savage and would always be grateful to him for he was the first to communicate to me the feel of the old community of which he was a sterling member'.[14]

Robert Savage was, according to Ivan, something of a gentle giant, a powerfully built man with a strong sense of humour, who would gently tease Priscilla his wife as he sat in his favourite fireside chair of an evening. The Savages had a good number of children, not all of whom lived beyond childhood. Ivan's father was born Robert Russell in 1911, but was always known as Russell. He was the only son to follow in his father's footsteps and become a shepherd. He'd also worked at Snape Maltings for a time, which, Ivan told me, only made malt in the winter months, when it was cooler, so his father had worked on a farm during the spring and summer. For most of his career, Russell Savage had been a shepherd on the Wentworths' Blackheath Estate.

Unlike most estate workers, who lived in a tied cottage, Russell Savage had a large house in Friston, a little more than a mile from Decoy Farm where the sheep lambed each spring. This house was where Ivan was born and, almost inevitably, he left school at fifteen and joined his father on the estate just as his forefathers had done in and around Blaxhall. His mother had been in service before she married, as had so many women down the generations, but Ivan was the last to follow the family tradition, at least for a few years.

Ivan's time at Blackheath was just a few years earlier than my own, so over several cups of coffee we talked about people and places we both remembered: tractor drivers Toby Ling and Reggie Driver; Willy Foster the estate odd-job man who drove an Austin A35 van, and Tony Kersey, who had married Willy's daughter and also drove a tractor. We laughed as we recalled how Tony, a rather proud man, had driven a Ford 4000, while Toby and Reggie had driven Ford 5000s, which were larger tractors. He always wore a tie, while the others dressed more casually. Toby's son Kenny, a lad of about my age, also drove a tractor, but sadly died a few years ago, as I discovered when I set out to find and interview him for this book.

Ivan's father had charge of two flocks of sheep at Blackheath. There were 150 pedigree Southdown ewes and four rams, which won many prizes at agricultural shows around the country, and 200 Cheviot ewes that ran with three Suffolk rams. The Cheviot–Suffolk cross is a popular lowland sheep, presenting few problems at lambing. They have a healthy appetite so grow faster than many other breeds. As a boy, Ivan would help his father take the sheep to various shows around the country. Other than that, he travelled little until years later, when he started his own business.

As well as an innate affinity with shepherding, Ivan had an interest in machinery, and at the age of twenty-one left the estate to work for a company that installed grain-handling equipment. He married and settled in Ipswich, later moving to a village a few miles outside the town. One thing led to another and in 1969 he set up his own business in the same field. The 1960s and 1970s were, he told me, a time of great change as everyone was moving from storing grain in hessian sacks to bulk handling, so there was a growing market for elevators and grain silos.

I remember seeing lorries, trains and ships loading and unloading sacks of grain along Ipswich docks in the early 1970s, when I walked through each day from the railway station to college. Ivan told me that, at about the same time, he'd had two men working almost permanently at Cranfields, one of those dockside mills, installing and commissioning new labour-saving equipment. There were still railway lines along the dock then; looking back, I realise that I was witnessing the end of an era, as until then grain had always been stored in sacks.

By the time he retired, Ivan employed more than a hundred men and was supplying equipment across the UK and into Europe. He told me how he would set off early in the morning, drive to Edinburgh to meet a customer, take them for lunch, seal a deal in the afternoon then drive back, arriving just in time for a pint at his local before closing time. It struck me that had he been born even fifty years earlier, he would probably have worked his whole life as a shepherd, occasionally travelling to Ipswich, but rarely any further. Ivan told me how, to mark his retirement, he'd taken his wife on a three-month world cruise aboard the *QE2* and how the family home had once belonged to Ipswich Town football manager Bobby Robson.

Despite his obvious success, Ivan Savage is a very modest man, who described the way his life had turned out as good luck, being in the grain-handling business at a time of great change, and so opportunity. He also told me he'd always made a point of hiring people who were better educated than him. He has three daughters, none of whom wanted to work with sheep, so the long line of shepherds has finally been broken. We met at The Hold, because one of his daughters, Mandy, works there as an education officer. She told me how thrilled she'd been, when studying for her history degree at the University of Suffolk, to discover that the 'Suffolk Lives' module had focused strongly on the writing of George Ewart Evans, who had written so much about her great grandparents, Robert and Priscilla Savage.

Greater social mobility is one of the benefits of living in the second half of the twentieth century. Ivan and Mandy, along perhaps with some readers, and certainly my own family, have had the opportunity to lead lives that three generations ago would have been unimaginable. Most of us have enjoyed greater material wealth, although the connection with place that once defined people has been lost. It's been claimed that most of us don't move far from where we were born, often remaining within a hundred miles. In my late sixties I chose to move back to the town where I grew up.

In many ways, Ivan Savage's life bridges the old and the new. He told me how in his teens, when the sheep were lambing, he and his father would alternate spending nights at Decoy Farm. Evans had described lambing as the shepherd's harvest; shepherds' wages were a little above those of other farm workers, and until the First World War Robert Savage was paid sixpence for every lamb that survived and was weaned. During the war years this rose to ninepence and, later, one shilling.

This encouraged good husbandry and compensated for the nights the shepherd would spend sleeping in a shepherd's hut out in the fields, although Russell and Ivan Savage had slept in one of the farm buildings at Decoy Farm, but I doubt it was any more comfortable. Most lambs are born early in the year, and the shepherd would have wanted to be on hand to help any ewe that got into difficulties giving birth.

The shepherd's huts were basic wooden structures, mounted on iron wheels so that they could be moved from field to field. Norwich firm Boulton and Paul built them with a stove and sleeping platform and may well have supplied the one that Robert Savage used. They advertised them at sixteen pounds and ten shilling at a time when a shepherd's wages were around eight shillings a week. On a Victorian farm they will have been a functional necessity, often accommodating the shepherd and his dog for days on end. Today, shepherd's huts are far more costly, sometimes providing fashionable garden offices for people who work from home.

When George Ewart Evans wrote *Ask the Fellows Who Cut the Hay*, he noted that 'at the time of writing there is not a flock of sheep left in the parish.'[15] This was a significant change, as sheep had played a vital role in maintaining the fertility of the light sandy heathland around Blaxhall for centuries. Evans recognised, as had Turnip Townshend centuries before, that the fertility of the light sandy land around Blaxhall was improved by the manure of generations of sheep. Robert Savage had trained two of his sons to be shepherds, but only one, Russell, had ended up working with sheep.

Sheep had played a key role in the evolution of British farming. Common land was 'the only form of farming in early times'.[16] Ancient villagers had collective rights to graze, take fuel and grow food, only producing what was needed within their community – that is, subsistence farming, rather than growing crops that could be traded or sold. Keeping large flocks of sheep, mainly for wool, was, explained Evans, 'the first big challenge to self-sufficient communities'.[17] They began producing more wool than they could use themselves, and this led to what Evans described as 'embryo capitalism'. The conflict between the clear benefits of keeping larger herds of sheep, with the limitations of available common-land grazing, was, as Evans observed, 'one of the main causes of the drive to enclose common land'.[18]

Cropped land was under the control of the lord of the manor and farmed in strips by his tenants; common land was that which could not be farmed in that way, for example heath and woodland. Each tenant farmer would have a number of strips of land under cultivation. You can still see land farmed in this way at Laxton in Nottinghamshire, where three large fields are shared and follow a strict three-year rotation of wheat, another cereal crop or beans, then fallow.

Throughout the eighteenth and nineteenth centuries, parliamentary enclosure acts brought together strips of land, with new boundaries being formed and often marked with hedges. This not only made farming more efficient, but created a rural landscape that existed until the 1980s, when tractors became much bigger, hedges were torn out and fields made much larger.

When I worked on farms, a twenty-acre field was large. Today arable fields are often considerably larger.

As a child I used to enjoy reading stories about farming. Illustrations of shepherds tending their flocks inevitably pictured the shepherd as a smiling old man, with a slight stoop, a white beard, and always wearing a smock. This left me with the slightly romantic notion that shepherds stood apart from other men on the farm, but I never really understood why, just as the ploughman always wore a jacket, the shepherd wore a smock.

The reason became clear when I read accounts of conversations George Ewart Evans had with Robert Savage. He called his shepherd's smock a slop and explained that because there were no openings back or front, the wind and rain could not get in and it was very warm. Evans described how the smock was made from a twilled linen called drabbet. The smock usually had a honeycombed pattern across the chest. Creating this pattern is an embroidery technique called smocking, which dates back 700 years and is still used today.

The smock had two inside pockets, which Savage told Evans were useful when you managed to snare a rabbit while out in the fields. Something that was never in my childhood storybooks was a picture of a shepherd with an umbrella, but on wet days Robert Savage would carry what he called a 'gig' umbrella (the modern golf umbrella is said to be derived from a gig umbrella). The combination of large umbrella and smock kept the shepherd dry in the fiercest of weather. Unlike the other farm workers, he could not leave his flock in bad weather to do indoor jobs in the farm buildings.

Robert Savage was perhaps one of the last shepherds to dress as shepherds had for generations. Further evidence of this was that he often carried a 'costrel', an earthenware bottle that would hang from a cord from the shepherd's belt. Costrels are more commonly associated with pilgrims, who, like the shepherd, had to carry their food and water with them as they trudged through the countryside. Today's shepherd will dress like any other farm worker and more often than not use a quad bike to move among his flock.

Today sheep wear ear tags that enable them to be identified when grazing on common land. Robert Savage used a more traditional method known as a bellwether, where one of the natural leaders in the flock would wear a bell round its neck. Sheep, like cattle, and perhaps people too, have a strict pecking order, with leaders and followers. Each flock had a different bell, so the shepherd could tell one from the other. Flocks of sheep tend to stick together and not mix with others, so this system worked. The practice of using sheep bells will have dated back to before land was enclosed.

Shepherds led a hard life, spending most of their time out in the fields, but perhaps mindful of their lambing bonus, they were also kind. Robert Savage usually had one or two what he called 'cossets', sheep he had hand-reared as lambs. Cosseting means to pamper or indulge, and this is what bottle-feeding an orphan lamb entails. These cossets would follow the shepherd wherever he went, seeing him as their mother.

In Suffolk, the shepherd traditionally led his flock, rather than driving it from behind with a dog. This was how biblical shepherds guided their flocks. A cosset, as something of a favourite, would wear a bell and follow the shepherd wherever he went. The rest of the flock followed the cosset, and his sheepdog would be behind the flock, making sure none were left behind.

While cereal crops produce one harvest a year, arguably the shepherd has two. He earns his harvest bonus for successfully rearing lambs, but also has sheep-shearing to deal with. Evans recorded an interview with Priscilla Savage's brother, Abraham Ling, born in 1887, who explained that 'the clipping season lasted about five or six weeks, in May and June'.[19]

Just as many farms took on additional labour for harvest, so too did they contract out shearing their sheep. The shearers would meet on a Sunday evening at the pub, the Blaxhall Ship (thought to have once been called the Sheep), 'to arrange where they would have to go during the following week'.[20] A team of nine or ten shearers could shear as many as 300 sheep in a day, starting at six in the morning and working until dusk. It was physically demanding work, shearing done by hand until mechanical shearing machines were invented by Frederick Wolseley, an Irish emigrant to Australia.

Wolseley patented his mechanical shears in 1884 and was later joined by an English engineer, Herbert Austin. They established the Wolseley Sheep Shearing Machine Company in London and transformed the job of the shearer. He could now work much faster and shear many more sheep in a day. The late nineteenth century was a time of great innovation, both at home and at work. The Wolseley company later diversified into building cars. I had not realised that my first car, a 1967 Austin 1100, owed its existence to a company that started out making sheep-shearing machines.

It was some time before mechanical sheep-shearing was adopted in Blaxhall. Hand shears were used well into the twentieth century. These have two blades, joined at a single or double bow handle, which meet and cut when you grasp them tightly, and spring apart when you relax your hand. Hand shears are still made today, and often used for 'dagging' sheep, necessary for cutting the soiled wool away from the sheep's rear end. This discourages flies from laying their eggs in the fleece, which hatch into maggots that then eat into the flesh of the sheep. Dogs and cats can reach round to keep themselves clean, but sheep cannot and so rely on the help of the shepherd.

A man might shear thirty sheep in a day with hand shears, but today's mechanical shears allow five times as many to be clipped in the same amount of time. Over a six-week shearing season, a modern sheep-shearer might handle more than 5,000 sheep. Shearers are usually paid on piecework, so much per sheep, so the faster a man can work, the more he will earn. Naturally he has to also take care not to nick the sheep in the process (in 2019 an Oxfordshire farmer set a new British record, shearing 784 lambs over a nine-hour day). I used to enjoy watching the sheep-shearing competitions at agricultural shows, where skilled shearers raced each other against the clock. It's not a job I have ever wanted to try; I much preferred working with cattle.

At the end of a day's shearing, the men would bind their shears shut by wrapping them with wool. Ivan Savage told me how this made them easier and safer to carry, but this practice had an added bonus. They would keep this wool in a bag, and at the end of the shearing season there would be enough for their mother or wife to spin and knit some new socks. Children would also collect the tufts of wool that caught on branches and brambles as the sheep passed, and this too would be kept until there was enough for their mother to spin. Little was wasted in those days.

Shearing sheep was thirsty work, and as Robert Savage's brother-in-law Abraham Ling told Evans, 'every five sheep they stopped for beer and gin.'[21] Fortunately for the sheep, the beer they drank was what was known as small beer, relatively weak, and home brewed. As Abraham Ling explained, 'sometimes one man brewed the beer for the company and they took a cask in the cart.'[22] The beer was drawn from the cask and drunk from 'a cup made from a sheep's horn that would travel better than a glass or china mug'.[23]

Shearing sheep with hand shears was hard work and things like repetitive strain injury not yet recognised. 'The wrist was very sore after the first day and all swollen up but they had to get used to it,'[24] Evans was told. The shearers then were paid five shillings for every twenty sheep they sheared, so if a man sheared thirty sheep, he would earn seven and sixpence for a day's work.

When you consider that at this time, just before the First World War, Robert Savage's wages were twelve shillings a week, you can see that, working six days a week, a shearer could earn four times as much in a week as the shepherd who looked after the flock. But the shearing season was just five weeks, and the shearers might not be regularly employed during the rest of the year. It was the opportunity to earn well that no doubt led Ivan Savage to return to the farms at shearing time for several years.

It was the tradition in Blaxhall, according to Priscilla Savage, that when a shepherd died he would be buried in his smock. This, she explained, was why Evans had been unable to find an example of a shepherd's smock in the village. It was done discreetly without telling the vicar. To bury a man in the uniform of his trade appears to be respectful, and indeed, another custom was to place a handful of wool in the coffin, 'to allow the dead shepherd to produce on the day of judgement, as an earnest that when on earth he followed that calling'.[25] This practice, Evans discovered, went back centuries, even before the Wool Act of 1666 prohibited burying all but plague victims or the destitute in anything other than a woollen shroud.

This legislation was passed to protect the wool trade that had since the Middle Ages seen East Anglian wool exported to Flanders. As Evans observed, 'here in East Anglia the fine churches, built one might say on wool, are living witness to the wealth from its trading'.[26] Many of the great churches of Norfolk and Suffolk were built with donations from men who had made their fortunes from the wool trade. The churches at Lavenham, Worsted and my own local church, Wymondham Abbey, are good examples.

The family of the deceased had to swear an affidavit to say that the corpse was clad in a woollen shroud, and if this was found not to be the case, a five-pound fine was levied. People resented this imposition, because as is the custom today, people wanted their loved ones buried in their finest clothes. It was not until 1814 that the Wool Act was finally repealed, although it had not been enforced for a few years.

In Blaxhall the Wool Act prompted the parson to start a new burial book in 1678 so that he could record that a woollen shroud had been used as the law dictated. Many East Anglian churches still have these volumes which became known as wool books. Such a law would not get passed today, but learning of it made me realise that I'd never thought about how my own parents and uncles were attired for their funerals. The undertaker had never raised the subject.

The importance of making sure a corpse was appropriately dressed was just one of a number of customs and traditions that Evans noted. Another custom that probably dates from pre-Reformation times was the practice of leaving a lit candle by the coffin, which was usually placed in the church on the evening before the funeral. One that was then left unattended drew criticism from one of the Blaxhall women, who said: 'A corpse should have a bit of company and a light – even if it's only a night light.'[27] Placing a lit candle by the coffin in church was supposed to keep the devil at bay.

As Tollund Man had shown me, wool has been woven to make clothing for a very long time. But interestingly, none of Evans's books refer to how the wool shorn from Blaxhall sheep was used. Instead, when writing about oral history, he mentioned the Rhodes family who were weavers at Saddleworth on the Yorkshire–Lancashire border. He was impressed that Lord Rhodes could name seven generations of family members, though for a member of the gentry this was perhaps far easier than for you or me.

But there will have been weavers in East Suffolk, just as there were weavers in Norfolk. My home in Norfolk was owned by a family of weavers from 1674 to 1740 and East Anglia exported wool to the Low Countries for centuries, supplying Flemish weavers. The wool trade was lucrative and vast fortunes were made by those exporting wool from Suffolk to the continent. Much was packed and taken by long trains of horses as far as Southampton or Dover, as well as being sent by ship from closer ports such as Mistley, between Ipswich and Colchester.

In the second half of the eighteenth century, the Industrial Revolution saw weaving become mechanised and move from being a cottage industry to an industrial process. Richard Arkwright is perhaps the best-remembered innovator of that time, developing a water-powered spinning frame. He was born in Lancashire and his weaving mills were built in the north of England, where fast-flowing rivers provided the power needed to drive the looms. The East India Company began to import cotton from the West Indies for these new mills at around the same time.

Perhaps knitting was such a universal activity in the early years of the twentieth century that Evans did not think it worth noting in his books. Producing garments from wool was, before the Industrial Revolution, labour intensive. It has been estimated that it would take a hundred hours to spin wool from a fleece and knit a single pullover.

I doubted if many people still had the time or patience to undertake such a task, but was mistaken. A visit to Galloper Sands, an intriguing arts centre based in ancient barns at White House Farm, Great Glemham, revealed a knitted ploughman's cape created by local textile artist Sarah Butters. She had taken a fleece from the farm, then carded, spun and knitted it into a rather impressive garment.

Keen to learn more, I contacted Sarah, who told me that she'd been knitting since childhood and loved working with wool and other natural fibres. We talked about how many jumpers today are made of synthetic fibres and made to hit a retail price point, rather than last. Sarah told me that a knitted woollen jumper could last for years, unlike those made of polyester. Polyester, as its name suggests, is a plastic, so far from environmentally friendly.

What really impressed me was that the colours with which the cape was dyed had been made from plants gathered on the farm: oak, willow, woad, mulberry, comfrey, nettles and cow parsley. I knew that tannin from oak bark was once used to tan leather, but learned from Sarah that tannin also helped dyed wool retain its colour. I realised that the colours associated today with country fashion – browns, greens and yellows – all were once derived from the very landscape they are chosen to represent. Dye-making was quite an industry, with a trade guild formed as long ago as the twelfth century.

Just as Suffolk had its own breed of working horse, the Suffolk Punch, so too did it develop its own breed of sheep. Suffolk sheep are easy to spot because they have black legs and faces. According to the National Sheep Association they are a large, fast-growing sheep 'that can be ready for market early, resulting in reduced input costs'. This is a polite way of saying they live short lives and so are profitable to keep. Unlike upland breeds, which are happy foraging on sparse moorland, the Suffolk sheep is more comfortable grazing grass and picking over cereal stubbles, fodder crops and, these days, the sugar beet tops left behind the harvester. I suspect that Blaxhall shepherd Robert Savage tended a flock of Suffolk sheep.

Keeping sheep that eat well and grow quickly comes at the price of convenience. While upland sheep can be left to fend for themselves, the sheep in Suffolk will have been far more intensively managed. You cannot let them have access to a whole field of lush grass or turnips, but instead have to erect hurdles or temporary fencing, or today electric fences, to let the sheep graze the field a strip at a time. This prevents waste, but more importantly stops the sheep over-eating as this can cause bloat.

Like cattle, sheep are ruminants and too much grass, especially in the spring, can cause them to produce more gas than they can pass. The swollen stomach presses up against the sheep's diaphragm, preventing it from breathing. Sheep with bloat die more quickly than cattle, which perhaps is why the old shepherds in Blaxhall carried a trochar. This was a hollow lance that when stabbed into the animal's paunch, and the centre withdrawn, would allow the gas to escape from the stomach. Thomas Hardy describes this in *Far from the Madding Crowd*; perhaps he too had seen this dramatic deflation of a bloated sheep and the instant relief it delivers.

The shepherd would use wooden hurdles to keep the sheep to one area of the field, moving them each day once the sheep had eaten all the grass or turnips. Hurdles were lightweight sections of wooden fencing, usually made from ash, hazel or chestnut, which is relatively light and splits easily. On large estates, the woodmen would coppice trees to provide a regular supply of three- or four-inch-diameter sticks. These would be split lengthways. Uprights would be drilled and cross pieces sharpened at each end to fit into the holes. Carefully hammered nails prevented the hurdles coming apart.

Ivan Savage told me that when he was working with his father on Blackheath Estate, one of his jobs was to take fifty-yard rolls of wire netting and erect it with a wooden stake every three yards, to keep the sheep to the areas of the field they were to graze. His father made hurdles, cutting lengths of elm from the woods on the estate. He would use the hurdles to make lambing pens in the fields. These gave the new born lambs somewhere to escape because there was a risk of the ewes lying on them in the first day or two of life.

WORK

The work we do largely dictates how we spend our adult lives. Like me, you may have moved to another part of the country, or even overseas, to follow your chosen career, yet it was not always like that. In rural England one hundred years ago sons expected literally to follow in their fathers' footsteps. Generations of boys and men worked on the same farms, and many joined the annual migration to Burton-on-Trent, helping turn barley into beer. Others followed traditional village crafts, repairing leather harness worn by Suffolk Punch horses as they worked the land.

We work to make money, but too often and certainly in my case, we define ourselves by the work we do. Work can consume all of our waking hours, often to the detriment of our family lives and even our mental health. But without work, we cannot afford to feed our families, house ourselves or take holidays. Work and life have for most become quite separate, often taking place many miles apart. I live on the Suffolk coast, a two-hour train ride from London, yet many of my neighbours make that journey to and from a highly paid city job every day.

When I met Daphne Gant, whose grandfather Robert Savage featured so prominently in George Ewart Evans's books, she told me how it was in the 1950s, when Evans lived in Blaxhall and was writing those books. Then she knew everyone who lived in the village, and would wave to the men who passed her cottage on their way to work the fields around the village. Of course there were people then who commuted, but this was rarely further than to the Bentwaters US air force base three miles down the road.

Recently there has been a move from office to home working, with broadband speed rather than rail connections defining where we base ourselves. Will this encourage us to once again get to know those who live around us and feel we are part of a community?

Leather

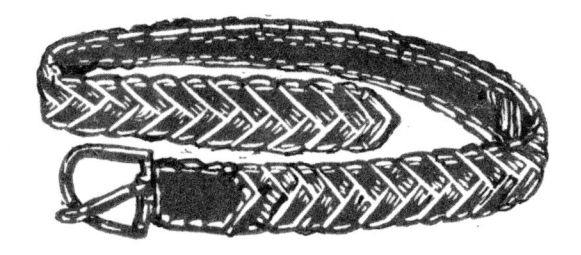

I VISITED A CRAFT FAIR staged in the grounds of Norfolk's Hockwold Hall one Sunday to help my wife, who had taken a stall to sell the polymer clay jewellery she makes. As soon as her gazebo and tables had been erected, I strolled across the parkland to buy a coffee and on the way found a stall selling felted sheep's-wool rugs. A banner across the front of the stall read 'Pastures Ewe' and as wool was still on my mind I went over to find out if the fleeces were locally sourced or imported, as so many products seem to be these days.

Jan Hannaford, the lady behind Pastures Ewe, told me that many of the fleeces came from a flock of sheep that graze on the parkland that surrounds a number of grand houses in south Suffolk. The fleeces, labelled with the breed of sheep from which they'd been shorn, had been bought from Portia Brown, who keeps a flock of 800 sheep, many of which were rare breeds, such as the Leicester Longwool, which owes its name to the length of the fleece. Knowing that Suffolk now had far fewer sheep than even in Evans's day, I contacted Portia and she kindly agreed to introduce me to her sheep.

We met by the war memorial in Clare's market square, an appropriately historic setting to begin an investigation into twenty-first-century shepherding, as Clare was at the heart of Suffolk's medieval wool trade. Portia, I learned, used to work in the City of London, in a sales and marketing role with a Japanese bank, but she'd grown up locally on her parent's smallholding and, now that she has three children, had decided to give them the chance to grow up in the countryside she had so loved. She'd started with a few sheep, then her flock quickly grew as people with parkland and water meadows offered her room to expand.

Today her sheep and cattle run across 600 acres she rents from a number of landowners. Some of the land Portia grazes belongs to people who bought it simply to shelter their wealth from inheritance tax, as farmland can be passed on without tax being levied. They are not farmers in the traditional sense, and I suspect they were pleased to have their permanent pasture managed by someone who, like them, was a first-generation farmer. Portia explained how she took care to keep fences well maintained and often put a few sheep onto paddocks that had been grazed by horses, as sheep crop the grass shorter, which does it good. I rather liked this symbiotic approach to estate management.

It was while standing in a field of Hereford cows and their calves that our conversation turned to leather. In common with many of the younger farmers I'd met, Portia was not content simply to accept what her sheep and cattle make at market. Instead she adds value, selling the fleeces I'd seen at Hockwold, and boxes of lamb, mutton and beef, to customers who value the fact that her animals are grass fed, well cared for and butchered locally. Portia plans to team up with a tannery, so that greater use can be made of the hides.

There is even a market, she told me, for rams' scrotums, which when tanned are used to make leather purses. Google revealed that in past centuries, these were popular as money or tobacco bags. The thought of using one myself makes me wince, but some of the old men I'd worked with on farms in the 1970s always referred to an animal's scrotum as a purse. I now understand why.

The 2020 pandemic had boosted demand for locally produced food and Portia had been kept busy meeting demand. While it would not be economical to set up her own small abattoir, she was planning to butcher her own meat on the farm where she and her partner, Warren, who runs a farm-contracting business, are based. She told me how she leaves before six in the morning when taking stock to the abattoir, so they can be processed as soon as they arrive and so suffer as little stress as possible.

My earliest memory of leather dates back to my first day at Needham Market primary school in September 1961. We'd just moved to the town from Brantham near Manningtree, where I'd been attending the village school for just three terms, and to soften the blow of making a fresh start at a new school, my mother had bought me my first pair of leather lace-up shoes. I'd been practising tying laces over the summer holiday.

Even now, the smell of new leather takes me back to that first day at Needham Market primary school. There is something about leather that says strength, maturity and, somehow, safety: a leather belt, a leather jacket and, in the times that George Ewart Evans wrote about, the leather harness worn by Suffolk Punches on the farms, or the leather belts that transmitted the drive from a steam tractor to the threshing machine working behind it. I often wonder if the horsemen Evans interviewed were similarly moved by the smell of leather as they harnessed up their horses before dawn each morning. As he wrote: 'The horseman was the earliest riser on the farm. He got up at 4 a.m.; took a bite of bread and cheese and hurried to the stable to feed the horses. Between their first bait (or meal) and their turning out to plough at 6.30 a.m. two hours must elapse. This was the unalterable rule in Suffolk.'[28]

There were many traditions – what Evans called unalterable rules – that dictated the way people behaved in the early years of the twentieth century. Speaking both Welsh and English, and having studied classics at university, he took a close interest in the words people used, as well as in what they said. He recognised Chaucerian words in the dialect his neighbours in Blaxhall spoke and realised that language and tradition here had changed little over the centuries.

He also recognised the way mechanisation of farming and the replacement of the horse by the tractor had broken down the insular self-sufficiency that had defined rural life for generations. As he wrote, 'The old pre-machine village community in Blaxhall was a tightly knit group, integrated for the carrying out of a particular work – the farming of the land and the many subsidiary trades and crafts directly connected with that farming.'[29]

Evans never tried to challenge or ridicule those ancient traditions, although as an outsider with a university education he would have been very aware that different ways of doing things could be equally if not more successful. Instead, he listened attentively, respecting both the stories he was told and their narrators. It was because he had won the trust of the people he lived among that he was able to record their memories and share them in the books he wrote. Grandchildren of those he interviewed told me that people felt flattered that he found their stories worthy of putting in a book. The centenary of his birth in 2009 was marked by a large, two-day gathering in a marquee next to Blaxhall's village hall, with a programme of talks, readings from his books and traditional folk music.

Too often, village newcomers will strive to impose their own version of a rural idyll, with little recognition of what already exists, or wistfully long for what they see as lost ways and traditions. Cambridge academic and contemporary of Evans, Raymond Williams, made the point in his book *The Country and the City* that for centuries writers have felt that they were describing a soon-to-be-lost age. He traces this back to Thomas More's *Utopia* published in 1516. There is little new about how people view the past.

Blaxhall is unusual in that its population has not grown much since the Middle Ages. Parish councillor Shane Pictor told me that today there are 123 houses in the village, with a population of around 250 people, while before the Black Death struck in 1348, there were probably 180 villagers. That pandemic killed just over half the population. It took more than a hundred years for the population to regrow.

I wondered if this was why the village was so spread out, rather than clustered around the church. I knew that some villages were abandoned at the time of the Black Death, with new settlements built a short way away, but village historian Rodney West told me that this was not the case. Blaxhall is what is termed a dispersed village, with the houses clustered around the various farms, rather than surrounding the church or village green. This is a reflection of how land ownership has evolved over the centuries, with many owner-occupied farms, rather than a single large landowner. The church, which dates back to the thirteenth century, stands alone close to a crossroads on what is called Stone Common.

Evans will have noted that Blaxhall had remained the same in so many ways for a very long time. He also realised that many of his neighbours were related, because they had the same family name. One of those he interviewed in the mid-1950s was Cyril Herring, who was postmaster in the adjoining village of Tunstall. He told Evans: 'A strange postman delivering letters in Blaxhall would find it an absolute bedlam to deliver among the forty Lings and twenty Smiths.'[30] Although when I visit Blaxhall today there are no Lings remaining, a good number of those I do meet are direct descendants of those old village families.

Evans understood that village life should be allowed to evolve naturally, without trying, as he put it, 'to preserve the old ways and customs artificially'.[31] This he described as 'misguided romanticism'. His son-in-law, the artist David Gentleman, in writing the foreword for the posthumously published *The Crooked Scythe*, explained: 'The underlying preoccupation in all George's books is with people and change. He knew that change, with its good aspects and bad, was inevitable, and he viewed it dispassionately.'[32]

Leather, and its evolution from a product that was made close to where it was used to a global, traded commodity, is a good example of the change that Evans was witnessing. Blaxhall's shoemaker made and repaired just about all the shoes worn in the village. People did not feel the need to travel in search of the latest fashion; shoes then were functional necessities. There was constant demand for the shoemaker's services, and he would have charged just enough to cover his costs and make a living. Today, shoes are mass-produced, distributed internationally and cleverly marketed to stimulate desire and deliver profit. They are no longer always repaired when they wear out, although there is a Timpson on most high streets. I might find the smell of new leather evocative of my childhood, but the leather I encounter today is unlikely to have been made within a thousand miles of where I encounter it.

That was not the case in the 1960s, when I had that pair of Clarks' lace-ups. The firm had started out in 1825 in Street, a Somerset village, by making sheepskin rugs using cull sheep from nearby Exmoor, where sheep had been grazed for centuries. Cyrus Clark, who founded the firm with his brother James, had earlier been apprenticed to a Somerset tannery and so inevitably they used locally sourced leather. Today, not only the leather, but the shoes themselves, are likely to have been made overseas.

Suffolk also has a long history of leather-making. The surviving lid of a sixth-century leather purse was found when the Sutton Hoo Anglo-Saxon burial ground a few miles from Blaxhall was excavated. One of Suffolk's longest-established businesses was the Webb & Son tannery, established at Combs near Stowmarket in 1711, and today in the ownership of a direct descendant of Joseph Webb, who took over the business in 1776. Nic Portway ran the business until cheap imports forced its closure in 1989. He still lives in the house next to the tannery that his ancestor occupied 250 years ago.

The tannery buildings today form a business park, occupied by, as Nic Portway told me, sixteen firms that between them employ more than 120 people. I was keen to learn a little about how leather was made in Suffolk, for in all probability the harness used by the horsemen Evans interviewed had been made from leather tanned at Combs. At its peak, he told me, the tannery had employed 250 people. There is an 800-acre farm as well, which has also been passed down the generations.

The tannery yard that Nic described, with a succession of lined pits dug into the ground around which the buildings were arranged, reminded me of a visit I'd made a few years previously to a camel tannery in the Moroccan city of Taroudant. This helped me imagine the incredible stench that will have made the tannery impossible to ignore. Hides from slaughtered cattle, sheep, pigs and even horses were scraped to remove hair and fat, salted to remove water and inhibit putrefaction, and then steeped in those pits, filled with water containing tannin leached from oak bark. Hides were moved sequentially from pit to pit, each containing a slightly stronger tannin solution. This preserved the hide, which when dried became leather. The process was dirty, smelly and very labour intensive. The workers' wives must have smelt their husbands coming up the garden path as they returned home from work each evening, stinking from the tannery yard.

Talking with Nic Portway reminded me of Evans's observation that a rural village will have been largely self-sufficient, with local tradespeople meeting local need. Combs tannery used hides from local livestock, oak bark stripped from trees at local sawmills, and water drawn from a well. The bones from the abattoir will have been ground to make fertiliser, used on the tannery farm and others before coprolites (fossilised dinosaur dung) were first mined and used to make more concentrated phosphate fertilisers. Nic also pointed out that until roads were improved in the early nineteenth century and the railway reached Stowmarket in 1846, transporting goods and livestock was a slow and at times dangerous activity.

Towards the end of the nineteenth century, leather from Combs tannery was exported worldwide, winning prizes at trade fairs in Egypt and even further afield. Many Suffolk companies prospered in the economic boom of the 1850s; manufacturing was relatively efficient and the British Empire provided ready markets for all that could be made. The expanding rail network made it easier to transport goods and many Suffolk companies enjoyed great success, making their owners very wealthy. Garretts were making steam engines at Leiston, Ransomes were making ploughs, barn machinery and threshing machines at Ipswich, and the Cobbold family's brewery and barley-shipping business was expanding fast on Ipswich docks.

But although Suffolk firms were enjoying international success, they never lost sight of the importance of their local market. Combs tannery supplied many of the village shoe- and harness-makers that Evans wrote about. One who almost certainly bought his leather from Combs was Sidney Austin, who was born in 1888 and lived and worked at Finningham, a village eight miles from the tannery.

When Evans interviewed him in 1960, he learned that Sidney Austin took over his father's business when he was a very young man and has kept it ever since, completing a period of about a hundred years since the business was first opened. He went on to note the decline of the firm from its heyday a few decades earlier: 'When horses were the sole power on the farm, he had five men working in his shop; today he has one.'[33] It must have been sad to see business dwindle as he grew older, although he continued repairing saddles well into his eighties.

Although Combs tannery was supplying leather just a few miles down the road, Austin's father would also take advantage of the opportunity to tan an occasional hide himself if a horse in the village needed to be slaughtered. A Suffolk Punch might have a twenty-five-year working life and be cared for by the same man for all of that time. The bond between horseman and his team was strong and intuitive, so it was natural that the unpleasant task of putting down a sick or old horse was left to someone outside the tight-knit community of workers on the farm. At Finningham, and I suspect elsewhere too, this unenviable task will often have fallen to the harness-maker, a trusted friend to the horsemen.

Sidney Austin explained this to Evans: 'They used to call us knackers years ago, because we used to dress our own leather: the horse slaughterer – knacker, as he's called today – and the harness maker were one business. The harness maker took the whole hide and tanned it himself.'[34] The harness-maker was in a good position to dispose of the animal and profit from the transaction. When I worked on Suffolk farms, the knacker man would arrive with a small truck, into which he would winch the dead animal and take it away to be rendered back at his yard. In the 1850s, he would have turned up with a horse, cart and a poleaxe, an iron spike on a long handle, which he would deftly swing to finish off the animal he had come to collect if it was not already dead

Another harness-maker interviewed by Evans enjoyed rapid promotion from apprentice saddler when his work colleagues were called up in 1914. Leonard Aldous, who was born in 1900, left school and started an apprenticeship at Debenham when still only twelve. 'My wages were sixpence a week. If I'd been fourteen, I should have commanded one shilling, but being only twelve and a half I got sixpence. Within two years of starting my apprenticeship, I was having to roughly do a man's work, but my wages were only half a crown a week,'[35] he told Evans

The importance of these local harness-makers cannot be underestimated. Repairing worn or broken harness was as important to the nineteenth-century farmer as servicing a van is to a delivery driver today. Leonard Aldous told Evans how each horse had his own harness, and because the horse had to work each day, the horseman would take the harness to Leonard Aldous's workshop at the end of one day and collect it in time to start the next. This shows the dedication these horsemen and Aldous had to their work, as he must have worked into the night to repair harness so that it was ready early the next morning.

As well as the horse being fitted with a leather harness, so too was it important for the horseman to be kitted out with strong boots. Evans described how horsemen of the 1880s would wear locally made boots, with an extra tongue to make them more comfortable as he would be wearing them for many hours each day. They were made by the village cobbler and cost more than a week's wages, so made to last. Walking behind a horse meant that a pair of boots would cover many miles, so each year they were what the horsemen would call 'clumped', which was to have a half-sole riveted to the bottom of the boot over the worn leather sole.

Style was more important for women's footwear, although

most also wore boots. Priscilla Savage, who was born in 1882 and lived next door to Blaxhall school, was one of those Evans interviewed in depth, largely because she was his neighbour and also, I suspect, because he was attentive and so she was more than happy to share her memories. It must have seemed strange that this Welshman living in the village was so interested in the everyday things that village folk took for granted.

She went into service at the age of eleven, as did so many girls of her generation (including my own grandmother). To mark this transition from dependent child to working girl, her father bought her a pair of boots, which, as she told Evans, meant something of a sacrifice as money was scarce. 'My father said he would get me a pair of boots – which of course were made by our village shoe maker, Mr Newson. He made me a pair of high kid topped button boots. I wore them for years and years.'[36] I suspect they will have been made a little on the large size, to allow for future growth.

People in late-nineteenth-century Blaxhall usually dressed up for their Sunday afternoon walk; it was the only day they did not work, set aside for church and leisure. Evans described how the off-duty horseman would have the outside of his trousers trimmed with horseshoe buttons, a decorative change from his heavy-duty work clothes. This would be complemented by a braid belt, as Albert Love told Evans: 'All the young horsemen used to wear them. It was the practice to walk out on a Sunday with a braided leather belt showing just below your waistcoat.'[37] These belts were often bought at Burton-on-Trent, by young men who, finding work on the farms had dried up after harvest, went to work on the maltings at Burton. They were souvenirs of sorts and also a kind of status symbol, rather like a designer label today. In the years just before the First World War the belts cost half a crown, according to Albert Love. Farm wages were then typically twenty-nine shillings a week; this was an investment of half a day's wages

I tried to think back to what I wore as a young man in the 1970s but don't think I was at all fashion-conscious. I did not go out to be seen in the same way as those horsemen had, more than 150 years earlier. When a working week was typically fifty hours or more, the Sunday afternoon walk will have been a highpoint; an opportunity to meet and talk with friends and neighbours. Walking was also how working people usually travelled from place to place, and to plough an acre meant to walk eleven miles behind a horse. To walk for pleasure, taking one's time and stopping to chat, will have been a treat.

In many ways the Sunday-afternoon walk was to the village horseman what 'going for a spin' in the car was to my father. I was nine when he bought his first car, a pastel-blue Austin A30. He took immense pride in driving us at the weekend from Needham Market to Kersey to feed the ducks on the River Stour.

Later, when I started work selling Fisons fertiliser to farmers in Norfolk, I joined North Elmham Young Farmers Club, which met a few miles from my then home. It was as a young farmer that I came to know the nearby rural life museum at Gressenhall.

The museum opened in 1975 and occupied what had originally been the village workhouse. Local farmer Dudley Crisp, a man with a strong interest in preserving the rural way of life, had played a key part in its establishment. His three sons, Richard, Anton and Mike, were all around my age and leading figures in North Elmham Young Farmers Club when I was a member. It was not possible to belong to the club and not visit Gressenhall, which with its collection of old farming implements was a fascinating place to explore. This all fuelled my interest in East Anglian farming history and the writing of George Ewart Evans.

It was my interest in leather, as I was researching this book, that prompted me to contact Gressenhall Museum again. I'd visited many times over the years, keen to share my fascination with farming's history with my children, who humoured me rather than taking any real interest. The museum was also a good place to take my father-in-law when he retired from farming. He missed his connection with the land, and the displays and old farming machinery at Gressenhall encouraged him to reminisce about his youth, when he had been the chaff boy working at the back of his grandfather's threshing machine during the war.

I wanted to meet a modern horseman and knew they kept some Suffolk Punches at Gressenhall. I also wanted to see leather harness, like those Evans had written about, and perhaps even experience walking behind a horse as it pulled a plough or cultivator through the soil. I emailed the museum and arranged to visit and talk with Richard Dalton, the farm manager.

The museum was quiet when I arrived, with lockdown meaning that visits had to be pre-booked and carefully timed. A coach in the car park that had brought a group of schoolchildren was a welcome reminder of the normality to which I hoped we would soon return. I was directed to the farm, a short walk from the museum. The museum has not always had a farm, but the tenant of the fifty-acre county-council-owned farm next door had retired, which allowed the museum to expand and show visitors how the land used to be worked in the years before the tractor.

Richard Dalton was a fascinating man. Approaching sixty, he had been running the farm for thirty-three years, yet remained endlessly enthusiastic about Suffolk horses and almost-forgotten traditional farming practices. He described how ploughing matches had changed, with few horses now taking part. 'The horses need to know what they're doing,' he explained, telling me that both horses and ploughman needed many hours working together to be comfortable with the task. When Richard attends ploughing matches, he takes his own plough, a

Ransomes YL, and the two horses he takes are twenty years old, so they know the ropes.

This familiarity was illustrated when I walked with Richard as he took one of the horses up the lane to a paddock, to graze with others. He put a halter over the horse's head, along with a fly mask to prevent flies clustering round his eyes. Standing beside a Suffolk horse, you become aware of just how big these animals are. Standing some seventeen hands high and weighing about a tonne, I realised that the halter and rope would be no help if the horse decided to go somewhere other than where Richard was taking him. But the rope was slack, for the horse knew just where he was going, and no doubt why.

We walked back to the farm and stood for a while chatting in the tack room, where a row of collars lined the wall and other harness was stored. Richard told me that a horseman will still describe putting the collar on the horse as yoking, a word that dates back to when oxen rather than horses were used to work the land. The collars are made of leather and filled with rye straw, which is tougher than either barley or wheat. It's the collar that distributes the load of what is being pulled evenly across the horse's shoulders.

Richard is also a trustee of the Suffolk Horse Society, an organisation dedicated to protecting this rare breed. There are fewer than 500 registered with the society, and considerable effort goes into matching the best stallion with each mare, reducing the dangers of inbreeding that are inevitable with such a small gene pool from which to draw. As with cattle, artificial insemination is more often than not used for breeding and, to my surprise, I learned that a Suffolk horse had also been successfully cloned.

A rare-breed preservation programme was launched by Shropshire-based Gemini Genetics. Their website explains how, as numbers of breeding males and females of a breed fall, the only way to build the gene pool and prevent inbreeding is to clone, to prevent the breed becoming extinct. Of the fourteen equine breeds native to the UK, twelve are considered rare and five, including the Suffolk, in a critical state, with fewer than 300 females.

Another surprise that afternoon was the discovery that Mike Crisp, whom I'd first known as a Young Farmer forty years ago, was the assistant manager and also regularly worked with the Suffolk horses. Mike had worked on the farm at Gressenhall Museum for more than thirty years and was now part time, preparing for retirement. His two brothers also farmed, both within a few miles of the village where the three of them were born and raised.

My career path had been more varied and not without its bumps, but looking back I rather envied the Crisp brothers for the way they had settled and remained close to the land. I, on the other hand, had felt driven to constantly keep moving and try new things. I might have been happier had I taken over my father-in-law's farm as he once thought I might, and become part of the Suffolk landscape that so fascinates me.

It was not by chance that the Gressenhall Museum farm kept Suffolk Punch horses. They were very much the draught horse of choice in East Anglia. First documented in the sixteenth century, the Suffolk was bred for its strength, stamina and docility. As a retired drayman at Adnams Southwold brewery once told me while showing me round, the Suffolk was bred for land work, and Shires and in his case Percherons, were larger horses, more suited to roadwork, so had been the breeds used to deliver beer around the town until 2010, when the last dray horse was retired. The stories those old horsemen told Evans will all have involved Suffolk Punch horses.

When researching his books on East Anglian life in the 1940s, 1950s and 1960s, George Ewart Evans interviewed the old people who lived near him or who came to his notice. So most of the people whose stories he recorded were born in the closing decades of the nineteenth century. They will have seen the change from the horse to the tractor, which meant that farms needed fewer men; those remaining spent more time working on their own than the old horsemen had. A pair of horses would plough an acre a day, while the Fordson tractors of the 1950s would manage that in less than an hour, and today's far larger tractors will plough more than four acres in an hour.

The retired horsemen that Evans interviewed will have grown up with one man and two horses ploughing an acre a day. 'A pair of horses had time to plough about 50 acres of land between harvest and the new crop being sown. Until the coming of the tractor, Suffolk farmers discussed the size of a farm in relation to the number of horse teams it supported, in much the same way as Domesday Book scribes measured farms in their survey of 1086.'[38]

A pair of horses would plough fifty acres between harvest and when the next crop was sown, so a hundred-acre farm would be described as a four-horse farm. Larger farms might have had a few older horses for carting and other lighter jobs. Arthur Young for example talked of a 250-acre farm being a twelve-horse farm, with five pairs for ploughing and two for odd jobs.

When these old men talked about their fathers' generation, they would describe how in the 1850s oxen as well as horses were used to pull the plough. George Rope (1846–1929) was born at Grove Farm, Blaxhall, and was Robert Savage's employer. Savage recalled how George Rope would talk about the oxen he remembered from his own childhood, explaining how after a few years' work, the bullock was grazed, then slaughtered, providing a good weight of beef. But a pair of horses could turn more tightly on the headland, so were more efficient and, over time, oxen were no longer used. No doubt he mourned the passing of the oxen as the following generation mourned the passing of the horse. I in turn am nostalgic for the early 1970s when I worked on farms and tractors were small enough to board without climbing a ladder.

Although times have inevitably changed, and farming continues to evolve, leather has remained in use throughout the centuries, growing today in popularity because it is more environmentally sound than plastics. As a material it is strong, durable and flexible, and although some will prefer to avoid it as it is derived from animals, it is arguably a more sustainable alternative to synthetic materials. As environmental concerns continue to grow, and people take a renewed interest in where things are made, leather is likely to remain important to us for many more years to come.

Barley

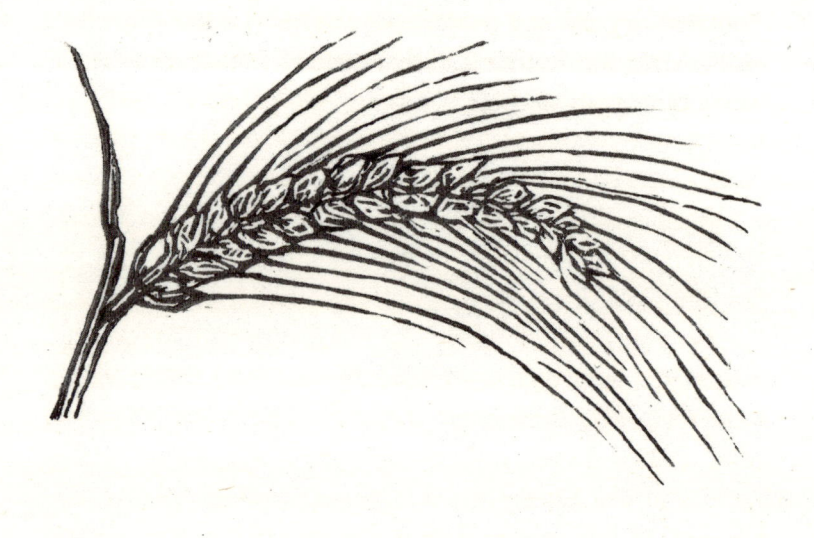

You can tell a field of barley from one of wheat by the whiskers, called awns, that stretch up from each ear and cause the barley to sway in even the slightest breeze. Barley is either crushed in a roller mill for cattle feed or processed to make malt, which in turn is used to make beer; wheat on the other hand is ground into flour to make bread. Both have been cultivated in Suffolk for more than 2,000 years and both were important to the people whose stories are told in the books of George Ewart Evans.

Evans's next-door neighbour in Blaxhall, Priscilla Savage, summed it up neatly when she talked of 'the three Bs – bread, bacon and beer',[39] which Evans described as being the 'staple articles of diet'. All were produced at home, as they had been for centuries. It was only the arrival of self-service supermarkets in the late 1950s and the cheap food they provided that prompted people to start buying, rather than producing, their food.

When Evans moved to the village in 1948 every cottage had a brick-lined bread oven that people still used to bake their own bread. It was common too for people to keep a pig at the bottom of the garden, and, until the early twentieth century, beer was also brewed at home, because being fermented it was safer to drink than well water, which was often contaminated.

My eighteenth birthday fell in August 1973 and, being a college student, I wasted little time developing a taste for beer. In East Suffolk at the time, there were two long-established local brewers, Adnams at Southwold and Tolly Cobbold in Ipswich. Both were family owned and both brewed good beer that was more than twice the strength of the small beer people used to drink at home decades earlier. I think it was at lunchtime on the last day of that autumn term at Ipswich College that I drank five pints of Tolly Cobbold bitter and managed to throw up on Platform 3 at Ipswich Station on my way home. I can remember the train guard insisting I rode to Saxmundham sitting on the floor in case I made a mess on the seats. A friend phoned my mother, who came over from Leiston to collect me. I woke the next morning with a memorable hangover.

We were lucky in those days to have two large local brewers, because through the late 1960s and early 1970s many local brewers were taken over by big national players such as Whitbread, Watneys and Courage. By 1980, 75 per cent of Britain's beer was brewed by just six companies, which between them also owned more than half the pubs in this country. To drive down costs, beers became weaker and, to reduce waste, most beer was pasteurised, which prevented further fermentation, and was delivered in stainless-steel kegs. Carbon dioxide was used to pump the beer to the tap, so the bar staff no longer pulled a pint in the traditional way. Even Tolly Cobbold tried to follow the trend, introducing Husky lager, which was inevitably likened to dog wee, because it was not very good.

An important ingredient of beer is malt, which is made by soaking barley, allowing it to germinate under controlled conditions, then heating it in a kiln to stop further germination. A tonne of malting barley makes three quarters of a tonne of malt, and, just as Ipswich had Tolly Cobbold brewery, the town was also home to R. W. Pauls' maltings. The Pauls are a well-known Suffolk family, farming on the Shotley peninsula and making animal feed as well as malt. Robert and Priscilla Savage's grandson Ivan introduced me to his friend Tony Bevan, who had spent his whole career working for Pauls', and then, when their malting business was sold to Associated British Malt, transferred from Ipswich to their far larger maltings in Bury St Edmunds.

Tony joined Pauls' as an apprentice electrician, helping maintain the six maltings the company had on and around Ipswich docks. Malting was then very labour intensive, with barley arriving in sixteen-stone hessian sacks and malt leaving in twelve-stone sacks. Barley was moved around the maltings by hand, too, from being steeped in water, poured onto the malting floor, where it had to be turned by hand every few hours, to finally being put into the kiln, from where it was bagged before being loaded onto lorries, ships or railway carriages – there were tracks laid along the dockside.

Pauls' had once owned a fleet of sailing barges and had been exporting malt world-wide since the nineteenth century. It had not occurred to me that barley cannot be grown in Africa or Japan, so malt had to be bought in from Europe, Australia or

North America. Every three weeks Pauls' would send a ship full of malt from Ipswich to Bremen, for the Beck's brewery. The ship would then bring steel from Bremen to Ford's plant at Dagenham, before running empty up the coast to Ipswich to collect more malt.

Talking with Tony was fascinating, as he had witnessed the malting business become increasingly more efficient over the decades, seeing small traditional maltings that were perhaps centuries old replaced by large, highly mechanised operations run by small teams. Felaw maltings on Ipswich's Stoke Quay had been built in 1904 and is the only maltings building now left in the town; it closed in the early 1980s and, being a rather attractive listed building, was converted into offices. In the late 1990s I often went there for business meetings, not giving a thought to its former industrial life.

Just as sacks were replaced by bulk handling, so too was the malting process progressively mechanised, with traditional floor malting replaced by Saladin boxes, which by mechanically raking the barley, and being able to cool then pneumatically force air through the barley, allowed malting to be carried out all year round. Previously it had been impossible to make malt during the summer as the ambient air temperature was usually too high.

Today, the whole malting process takes place in large steel drums that slowly rotate, keeping the barley moving during the vital germination stage. The barley is steeped in water and then allowed to germinate, blown through with cold air, and finally at the kilning stage, blown through with hot air in the drum. My simplistic mind likens these advances to the way domestic laundry has been automated over the same half-century. I compare the Saladin box to the twin tub I can remember my mother using, and drum malting to the modern washing machine that, once loaded, washes, rinses and often even dries each batch of laundry.

When Tony entered the malting business, barley was malted in twenty-tonne batches, and when he retired the maltings where he worked at Bury St Edmunds was producing 175,000 tonnes annually. Today the Bury St Edmunds maltings is owned by Boormalt, a Belgian corporate that is making 3 million tonnes of malt a year at twenty-seven plants around the world. But while malt is now a globally traded commodity, there is growing interest in craft beers brewed locally, often with malt made from Maris Otter barley.

Shepherd's wife Priscilla Savage brewed beer in her kitchen. Like most of her generation, she would do two large brews a year, once just before harvest and again in December before Christmas. She and other cottagers had an array of wooden beer-brewing equipment and four-and-a-half-gallon wooden barrels called pins, in which they stored the beer. To make a single barrel of beer she would buy 'one bushel of malt costing 14 shillings; one pound of hops worth about one shilling; one pint of yeast passed on from a neighbour's brew'.[40] The malt would have been bought from Snape Maltings two miles away, with only the hops having travelled further, probably coming from Kent to the village shop.

When beer was being brewed, other household chores had to be left undone or, as was more often the case, 'the children were kept home from school to do odd jobs about the house'.[41] No excuses were made for these absences from school, as Evans found when he examined the old log book at Blaxhall school. An entry from early August 1871 noted the reason for many children not being at school was because they were helping their mothers brew beer. Attendance at school was not compulsory until the 1880 Education Act made it mandatory to attend school between the ages of five and ten.

Beer was also brewed on the farms, with one of the horsemen usually in charge. This was done on a larger scale than in Priscilla Savage's kitchen, with hogshead barrels being filled to provide enough beer for all the men on the farm. As with most farm jobs, brewing beer was quite labour intensive. These barrels held fifty-five gallons and stood four feet tall. They had to be cleaned out and prepared before each brewing; Evans noted how the back-house boy, who was the most junior and perhaps often the smallest servant, would be asked to climb into the barrel, scrape it clean and then limewash it, ready for the new brew. These barrels were so large that the brewer would 'test the air with a candle', although I suspect this was more about teasing the boy than being due to any risk. Sitting in a large barrel is not quite the same as being lowered down a well or mine shaft.

There was a time in the 1970s when pub beers were less appealing than they were before or indeed are today. Price not quality became the driver, and cask beer, served from a hand pump, was largely replaced by keg beer that had a longer life. This also made pub work easier, as far less skill was required to change an empty keg or pull a pint.

Growing barley today is relatively straightforward. Farmers now have large tractors to prepare the seedbed, precision drills to plant the crop, chemicals to control weeds and efficient combine harvesters that quickly convert a standing crop into grain and chopped straw. In the 1890s, when those Evans interviewed were young, it was markedly different. The field would have been ploughed in the autumn by horses to turn over the soil, burying weeds and debris from the previous crop, then broken down to a fine tilth with a horse-drawn toothed harrow. Hand sowing and weed control was the norm, with harvesting also done by hand. This was backbreaking work, which is why the farmer provided his workers with plenty of beer.

Most farmers take real pride in their work. My father-in-law Michael would always prefer to be driven in his car than take the wheel himself. His reasoning was simple: he wanted to spend the journey looking over hedges to see how his neighbour's crops were doing. Strips missed by the seed drill or sprayer were easy to spot and quickly became talking points whenever farmers got together, in the pub, at the cattle market or an agricultural show. Marrying into a farming family gave me plenty of opportunity to listen to the good-humoured criticism farmers will level at the mistakes made by their neighbours.

The old men that Evans interviewed, although they were more often farm workers than actual farmers, also took a pride in their work. Their grandfathers had probably worked the same fields and had every expectation that their grandsons would do the same. Reputation was important; nobody wanted to earn a place in local folklore for being unable to plough a straight furrow. 'The care given to the sowing of the barley by the East Anglian farm worker was', wrote Evans, 'almost akin to veneration. No trouble was spared to ensure the carefully prepared seedbed was ready to take the crop.'[42]

The modern farmer has weather forecasts and soil thermometers to help him decide when to sow his barley. The old men Evans interviewed did not have technology, but were very aware of the importance of sowing their crops as soon as the soil conditions allowed. I'm sure Evans was smiling when he wrote of one of his interviewees that, to test the warmth of the soil, 'he took off his trousers and sat down on the seedbed, thus testing the warmth of Mother Earth with the most sensitive part of his anatomy.'[43] Evans described this more politely than others I have read, who described it as the 'bare bum test', which apparently is particularly popular even today among some potato growers; I just hope they don't have roadside fields.

Sowing the crop was for centuries done by hand, the seed barley carried in an open hod in front of you, suspended by leather straps that went around behind your neck. The skilled sower would then walk slowly across the field, throwing a handful to the left and then one to the right. There was a skill to spreading the seed evenly, and not applying too much or too little. Then the field would be harrowed again to work the seed into the top inch or so of the soil. When you consider that an acre of barley required almost a man's own weight in seed, you can see that sowing by hand was not an easy task.

By the middle of the nineteenth century brothers James and Jonathan Smythe were making horse-drawn seed drills at Peasenhall. These revolutionised sowing and the seed drills used today have evolved from those early models. The seed was placed in a hopper that ran the full width of the machine, then as the horse, or more usually two horses, pulled the drill across the field, a series of gears driven by the land wheels rotated small cups that picked the seeds up and dropped them down a tube, at the bottom of which was a coulter that opened up the soil.

The coulter was pointed, rather like the tip of a plough, with the seed dropping down the back of it into the opening it made in the soil. Weights were attached to prevent the coulter riding along on the surface. More weight was needed on heavier clay soils, heavy land being harder to penetrate. A series of gears allowed the seed rate to be adjusted, by speeding up or slowing down the rotating cups. The Smythe drill was in its day as revolutionary as the GPS-guided driverless tractors that are being developed today, both designed to save time and labour.

One job that has almost completely disappeared from farms is stone picking. Once commonplace, it now rarely takes place unless potatoes are being grown, in which case the stones are mechanically moved and laid between the baulks into which the potatoes are planted. In the nineteenth century, stone picking was thought to increase the yield of wheat and barley and, as labour was plentiful and cheap, stone picking was a regular task in the farming year. Stones taken from the fields were used to mend the roads, until 1888, when county councils became responsible for road maintenance. Before then each parish had to maintain its own roads; there were no quarries near Blaxhall, so using stones picked from the fields was the easiest way to repair the highway.

Stone picking was a family affair as it was a good way for the women and children to earn some money, although it was not a task anyone seemed to enjoy. Evans discovered that old memories of stone picking still remained in the early 1950s when he lived in Blaxhall. Having been a prize-winning sprinter in his youth, as well as having taught PE, he had suggested clearing a patch of common land at Blaxhall to create a playing field where he could coach some of the village children. But preparing the playing field meant clearing it of stones, and, as he wrote, 'Many middle aged people remember the time they spent stone picking with distaste.'[44]

Picking stones from barley fields was done after the newly sown crop had emerged and was a few inches high. The farm men would first rake the field to loosen the stones before the woman and children picked them up and piled them at the side of the field. You might think that having people walking all over a recently emerged crop would do it harm, but today barley is usually rolled after the first few leaves have emerged, to encourage the plants to produce more shoots, called tillers. As each shoot later bears an ear, this can increase the yield. Evans explained this more simply: 'Treading down the young corn was said to benefit it.'[45]

Trampling down the barley was a beneficial side effect of stone picking, and inevitable as the women and children criss-crossed the field collecting stones in two-gallon pails, which they then tipped in a heap. Eighty pails of stones were considered a cart load, and these heaps were left on the field until after harvest, when they would be carried away with a horse and cart. The farmer would inspect each heap and after harvest, when the stones were collected, pay the pickers three shillings for each cartload. 'He could fairly accurately tell how much each family had picked and he was accustomed to give the mother a little money on account.'[46]

The Smythe drill planted the seed in rows six inches apart across the field, while broadcasting by hand meant distribution was random. So it became possible in the late nineteenth century to hoe a field of drilled barley to remove weeds growing between the rows. This was mechanised, in that a line of hoes mounted on a frame were pulled across the field by a horse. Retired horseman William Cobbold, born in 1883, described to Evans how this was done: 'When the corn was about two inches high we hoed it with a horse hoe. This had nine whole hoes and two half hoes – one at each end.'[47] The half hoe would hoe half of the outside row when you went across the field, and the other half as you came back. Both horseman and his horse must have become skilled at following the rows of barley.

It was as recently as 1945 that the first selective herbicide, called 2,4-D, was developed by ICI. It was called selective because it killed broadleaved weeds but left grasses, of which barley is one, untouched. This was useful to a point, but one of the main weed problems with barley is black-grass, which grows vigorously, competing with the crop for nutrients and water and so suppressing yield. Today's herbicides are selective to the point that they kill both broadleaved weeds and black-grass, but leave barley or wheat untouched. The organic farmer will rely on frequent cultivation of the soil before the barley is sown, and lightly harrow the growing crop to control weeds that emerge later. But nobody today controls weeds in barley with the precision of the late-Victorian farm worker.

Barley yields have increased significantly over the past 120 years. Those who shared their memories with Evans will have seen yields double from a little over half a tonne per acre in the late 1890s to more than a tonne in the 1950s when he recorded his interviews. Today it is not uncommon for a crop of winter barley to yield almost four tonnes per acre, although Maris Otter, considered the very best malting barley variety, might yield half a tonne less. Maris Otter is prized among craft brewers and was bred in the mid-1960s at the Plant Breeding Institute near Cambridge. The variety is now co-owned by a Norfolk agricultural merchant, H. Banham Ltd, and is still grown in East Anglia, where the light soils and dry climate are perfect for growing malting barley.

To survive in the modern age, an agricultural merchant has to specialise, and many long-established family-owned merchants have been swallowed up by global grain traders over the last fifty years. When I was selling to farmers in the late 1970s, Norfolk had a number of agricultural merchants: Myhills at Wymondham; Stimpson Pertwee at Reepham; Tyrrell Byford & Pallet at Attleborough; Banham at Hempton; and Burroughes at Bressingham. All but Banham have now disappeared, although Myhills remains as a small chain of pet and garden stores, with outlets in several South Norfolk market towns.

To gain an insight into how these firms all disappeared I met John Burroughes, who had been an agronomist advising farmers on husbandry for many years after joining the firm founded by his grandfather. Burroughes's family had been millers in Norfolk since the nineteenth century, diversifying into animal feed before becoming agricultural merchants. Soon they were also selling seed, then fertilisers and agrochemicals as they were developed in the 1950s and 1960s.

Burroughes's father Eric would set off most days to the corn-halls to buy barley and wheat as his company had key supply contracts with maltsters, brewers and millers. On Monday he would visit Mark Lane corn exchange in London, Tuesday it was Ipswich, Wednesday Bury St Edmunds, Thursday might see him at King's Lynn, Friday was his local market at Diss, and on Saturday it was Norwich. At each he would visit the various traders' desks, where samples were examined by eye, with grains cut to assess the colour and texture inside. Deals were sealed with a handshake, because trust was valued more than a written contract.

The establishment of the European common market, which Britain joined in 1973, meant more protection for farmers, with predicted grain surpluses stored to maintain market prices. This created a new market opportunity for those in the grain trade, and because the government paid for what became known as 'intervention storage', and because Eric Burroughes had an interest in haulage, building up a fleet of distinctive lorries, painted green with contrasting red lettering on the cab doors, it was decided to borrow money and build a huge grain store on the site of the original family-owned windmill.

Unfortunately, government funding for grain storage ceased before the bank loan had been repaid, and this put pressure on the business. At around the same time, American grain-trading giants were pushing into the UK market, with firms such as Conagra, Bunge and Cargill buying family-owned merchant businesses to build their market share. These newcomers were buying grain from the farm and taking it straight to ships at King's Lynn and Ipswich. What had been for decades a gentlemanly trade, conducted in farmhouse kitchens and corn-halls, became a cut-throat business in which the Burroughes family found it increasingly difficult to compete.

When the business ultimately failed, Conagra bought the premises, took on some of the staff and acquired the connection with local farmers they had been seeking. Further consolidation and change within the grain trade means that today the Bressingham operation forms part of Openfield, a Lincolnshire-based grain-trading company with an annual turnover of more than £600 million. Calling on farms, as I once did, now seems as dated as those old corn-halls. Today, it's all done online.

Much of the increase in barley yields over the past century or so is due to the introduction of nitrogen fertiliser. Cereals cannot absorb nitrogen from the air; rather, it is taken up by the roots, having been converted by soil bacteria into a form the plant can use. Nitrogen is an essential element for the creation of chlorophyll, so providing additional nitrogen boosts the plant's capacity for photosynthesis and this in turn increases the grain yield. The industrial process by which nitrogen is taken from the air and blended with hydrogen from natural gas converted to ammonia, from which nitrogen fertilisers are made, is called the Haber process, and was first developed in Germany in 1910. Ammonium sulphate was the first nitrogen fertiliser used in Britain, with ammonium nitrate, a more concentrated product, introduced in 1965.

Fourteen years later, I joined Suffolk-based fertiliser manufacturer as a sales rep, and so made my living selling both ammonium nitrate and compound fertilisers that also contained potash and phosphate. My last two years with the firm were spent in the marketing department, working with an advertising agency to promote the 600,000 tonnes of nitrogen fertiliser the company made each year at its Immingham plant.

The fertilisers spread on the land at Blaxhall in the 1950s will have been delivered as powder in sacks. Horse-drawn fertiliser spreaders were similar to the Smythe drill, with a full-width hopper and adjustable gears driven by a chain running from the axle to allow the horseman to adjust the application rate. Fertiliser spreading will have been a dusty job, as it was not until the 1960s that fertilisers were granulated or, like lead shot, formed into small smooth round balls by dropping them down tall cooling towers. This was how ammonium nitrate was made; the tiny balls were called prills.

The advent of fertilisers was not as readily accepted by Blaxhall farmers as you might expect. Evans wrote about Robert Sherwood, who was born in 1885 and farmed at Blaxhall. He preferred to rely on meticulous record keeping to judge when best to plant, harvest and even sell his crops. As Evans observed:

'The new farmer – many of the old school maintain – no longer relies on his field-books but on his fertilisers; and he expects these faithfully to cover up his mistakes and bolster his shortcomings.'[48] It seems some believed that to use fertiliser was somehow to cheat at farming.

But then men and horses had been working in perfect harmony with nature for generations and fertilisers, like herbicides, although beneficial, must have seemed unnatural, almost magic in their effect. As Evans went on to write: 'Cultivation is no longer by rule or rote, but by formula; and the modern fashion is to ignore the living quality of the land and look upon it as a rather complicated chemical whose rhythms can be effectively controlled from the laboratory.'[49] An old farm worker offered a word of warning about what to many today will appear to simply be progress: 'In time they won't grow as much corn on the land as they're doing today. Chemicals are not putting back into the land everything the crops are taking out.'[50]

Perhaps he had heard of the Haughley experiment, where in 1939, ten or so years before that farm worker's conversation with Evans, two neighbouring farmers near Stowmarket ran field trials to compare farming with fertilisers, with what we would describe today as organic farming. One of the farmers, Lady Eve Balfour, speaking at an international conference on organic farming in 1977, suggested that over more than thirty years, she had on her farm 'clearly demonstrated that its fields had become dependent on their fertiliser supplements in a manner suggestive of drug addiction'.[51] Eve Balfour was one of the founders of the Soil Association, a charity that seeks to encourage a return from intensive production to a system that is more nature-friendly. According to the UK government, just 2.7 per cent of the farmland in Britain today is farmed organically; almost all agricultural land receives fertiliser.

In most years harvest time starts in July, with the combine harvester progressing on most farms from oilseed rape to barley and finally wheat. It is only in the past forty-five years that oil-seed rape has become widely grown in the UK. In the 1890s it was the hay that was harvested immediately before barley. Back then, far more hay was grown than today, being needed to feed horses, both on the farms and in the towns and cities. It is estimated that in 1894 London alone had 50,000 horses engaged in transporting people around the city. As Evans wrote: 'After the haysel (the time of the hay harvest) came the barleysel, the harvesting of the barley, which was the typical corn crop grown in East Anglia in Anglo-Saxon times, as it is in parts of the region today.'[52]

Wet weather can slow the harvest down, as barley will not store well if harvested when the moisture content is greater than 14 per cent. In the early 1970s, when I worked at Blackheath Estate, I would often spend weeks at harvest time in the grain store, where grain arriving from the field was tipped through a steel grating into a deep, concrete-lined pit. From there elevators and conveyors would carry it to silos that between them could hold more than 1,000 tonnes of grain. One of my tasks was to monitor the moisture content of the grain arriving at the store. This was done by grinding a handful of the barley or wheat from every few loads as they arrived, then putting it in a small electronic machine that gave a reading of the moisture content.

The old men Evans interviewed would have tested grain moisture in the same way my father-in-law Michael did, as he never bought a moisture meter. He would simply rub a few ears of barley between his hands to remove the chaff and awns, and then bite some grains between his teeth. The easier the grain was to bite, the higher the moisture content, so with experience it was possible to tell when the crop was ready to harvest. He will have learned to do this from his father and grandfather, both of them farmers. Michael only grew about a hundred acres of barley, much of which he would store in a small hessian-sided silo, later milling it himself to feed his pigs. Accurate measurement was less important than it was on a large estate, where a single silo might hold more than 200 tonnes.

Before mechanisation in the late nineteenth century, barley was cut with a scythe and, according to Arthur Young in his *General View of Suffolk*, written in 1797, raked by women into rows and carted loose back to the barn. Wheat on the other hand was tied into sheaves before being carted away from the field. Both were stored in a barn that more often than not had two large doors high enough to admit a horse and fully laden wagon on one side, and a lower pair on the other side through which the horse and empty wagon would leave. The area where the horse stood was also where the corn would later be threshed with flails on the floor, separating the grain from the straw.

When the barn was being filled, 'the quietest horse on the farm would be walked round and round to trample the corn down as it was unloaded. Then when it was as high as it could be, the horse would be lowered on a rope from a beam in the centre of the barn'.[53] For many years, I lived in a nineteenth-century brick-built barn spanned by four large beams roughly hewn from tree trunks. Having discovered how a horse would once have been lowered from one of the innermost beams, I couldn't help but picture this in my mind's eye whenever I glanced up from my dinner.

It would be wrong to assume that life on a Victorian farm was an unchanging rural idyll. The pace of change through the second half of the nineteenth century was rapid, perhaps more so than over my lifetime, when tractors, farm machinery and fields just became bigger. The invention of the horse-drawn reaper-binder in 1872, which mechanically cut the corn and bundled it into sheaves, dramatically reduced the labour needed at harvest time. First developed in the USA, Albion reaper-binders quickly became common across the UK, being made by Lancashire-based company Harrison, McGregor & Co. Ltd.

Like the Smythe drill before it, the reaper-binder revolutionised farming.

Previously, corn had for centuries been cut by hand with a sickle or scythe by the men on the farm, and even after the reaper-binders arrived, the headlands would still be cut by hand to reduce wastage. Although farmers may have welcomed the coming of the reaper-binder, farm workers were not always so enthusiastic. It was 1911 before any kind of unemployment benefit was introduced, and to see your job done by machine could mean finding yourself without work. It was no surprise that some took to breaking these machines, just as the Luddites had earlier opposed the introduction of mechanisation to the weaving industry.

Another nineteenth-century innovation was the threshing machine, which, once steam-powered traction engines appeared in the 1850s, could be much larger, driven by a leather belt from the powerful engine. This replaced threshing with a hand-held flail on the barn floor and remained the way barley and wheat were separated from the chaff and straw until the combine harvester became common after the Second World War. Until then, hand threshing had been monotonous, dusty work and those Evans interviewed said it was lonely too. 'Time passed slowly, and to mark the passing hours the men cut notches in the side of the barn'[54] that would catch the sun at different times, marking the passing hours.

My wife's great-grandfather William Easy had a threshing machine and Garrett steam tractor that he would use to thresh his and other farmers' corn in the villages near Theberton where he farmed. Her father Michael remembers well how hot, dusty and dirty it was being chaff boy in the 1940s. His own father, David, had taken a farm at nearby Leiston and it was here that, as a boy, Michael most often worked, beside the noisy machine pulling chaff away. Michael was born in 1938, so was one of the last to work with a threshing machine as the combine harvester largely replaced it. The chaff boy might have been the junior member of the team, but if he neglected his job the threshing machine would quickly block up, and threshing gangs were paid on results, not by the hour, so any delay was costly.

My own first memories of harvest time are as a fifteen-year-old, standing on the back of a Massey Harris combine harvester that filled sacks, rather than a tank with the harvested grain. My job was to put the sacks on the grain spout, tie them when full and slide them down a chute to the ground. A sack of barley would weigh twelve stone and a sack of wheat sixteen, which I really struggled to manhandle. I was soon told I was too puny and moved to baling and carting straw, which I found much easier. Today I own a Fordson tractor, identical to the one I drove that summer. Driving it always takes me back to the barley harvests of the early 1970s.

Coins

From the age of seventeen until shortly before his death, my father worked for Barclays Bank. He instilled in me the importance of saving and one of my earliest memories is of the blue china piggy bank he bought me when I was five years old. Its arrival coincided with my first week's pocket money, together with encouragement to save a few pennies each week. To this day I find saving easier than spending and worry unnecessarily about having enough money put away for my old age.

George Ewart Evans's relationship with money was also shaped by his father, who had what he described as a 'crippled right foot' and had been apprenticed to a grocer, before opening a shop himself in Abercynon, a mining village near

Pontypridd. Describing his childhood in his autobiography, Evans wrote: 'At that time I was sleeping in my father's bedroom. In the corner of the room was a heavy safe where he kept his deeds and valuables. Alongside the safe was a chest of drawers and in the second drawer from the top was a policeman's truncheon.'[55] He went on to describe how his father had told him that when as a 'youth he had been round the farms and cottages collecting accounts, he had been set upon and robbed of his master's money.'[56] George will have grown up knowing that money was to be respected, and protected from those who might try to take it away from you.

When I was a schoolboy, small change was always referred to as coppers and comprised pennies and halfpennies. Farthings, of which there were four to a penny, and so 960 to a pound, were withdrawn from circulation in 1960. It is hard to imagine now, looking back, that you could buy anything with a coin of so low a value, but one pound in 1960 had the spending power of £23 today. Now you only really see a penny, which since decimalisation in 1971 is just a hundredth of a pound, when you have bought something in a shop that is priced at say £9.99, handed over a £10 note and been given a penny change.

The change to decimal currency was, as I recall, not welcomed by everyone. I'd grown up knowing that there were twelve pennies in a shilling and twenty shillings in a pound. When I was at school we still had half-crown coins, worth two shillings and sixpence, even though the crown itself was long gone. The new decimal coins were smaller and appeared to be worth less. Some old coins remain, in our language if not in reality; for example the guinea, named after the West African country Guinea, where much of the gold used to make it was mined, remains in use in the horse-racing world, despite the last guinea coin having been struck in 1813. A guinea was worth one pound and one shilling, or one pound and five pence today.

There are other hangovers from our old currency. The scouting movement first introduced 'Bob-a-Job Week' in 1949, and reintroduced it in 2011 to encourage young people to help their local community. A bob, for reasons nobody seems to remember, was a popular name for a one-shilling coin, which in today's money is five pence. Older people still talk about spending a penny because to gain access to a cubicle in a public toilet before 1971 meant dropping an old penny into a slot in the door to release the catch. Our language is peppered with reminders of our pre-decimal money.

Blaxhall shepherd Robert Savage was born in 1880 and started work at the age of twelve. He worked with sheep all his life, and was paid a little more than a general farm worker. But this was still not much, as Evans found: 'Robert Savage's wages when he started with the sheep was eleven shillings a week. This was a little above the ordinary farm-worker's wage and fixed by yearly agreement.'[57] When he married in 1902, his pay rose to twelve shillings a week (equivalent to £75 today) and on top of that, he received a six-penny bonus for every lamb he successfully reared.

To get a sense of what twelve shillings would buy in the early years of the twentieth century, I consulted my copy of the 1909 edition of *Mrs Beaton's Book of Household Management*. Written in London for a middle-class reader, prices listed will have been higher than in Blaxhall, but as the preface states, the figures quoted had been 'carefully, minutely and diligently averaged from the most reliable authorities all over the kingdom'. Butter is listed as costing one shilling for a pound and chickens two shillings each.[58] It is no wonder that the people living in rural Suffolk made their own butter and kept chickens, as well as that pig in their garden. Mrs Beaton's book recommended paying a parlour maid between twelve and twenty pounds a year, then a few pages later lists Bollinger champagne as costing thirty shillings for a case of twelve bottles. Even allowing for inflation, which in the early years of the twentieth century averaged just half a per cent a year, this illustrates well the chasm between what was affordable by the farm worker and what the large landowners that employed them were able to buy.

Rural poverty was the norm throughout most of the nineteenth and early twentieth centuries. When I met Daphne Gant, a granddaughter of Robert Savage, who has lived her whole life in Blaxhall, she told me that before she retired, she had always taken on what work she could find to supplement her husband John's farm-worker wages. She said that they'd not opened a bank account until their daughter had left home and encouraged them to do so. John had been paid in cash at the end of each week, as would Daphne for any work she had done. His pay covered the rent of their cottage and household costs. What Daphne earned was saved up to buy things they could otherwise not afford, for example a fridge. But although to an outsider this way of living hand to mouth may appear harsh and austere, they seemed to me a contented couple with no sense of having missed out on things they could not afford.

It was not just the farm workers who had to manage on a meagre wage. While the years following the 1850 Great Exhibition had been prosperous, with farm profits growing on average by 24 per cent over two decades, the years between 1875 and 1896 were termed the Great Depression. Wheat prices had plummeted in response to cheap imports from the USA, falling from six pounds and eight shillings a tonne in 1877 to less than five pounds a tonne in 1895. Tenant farmers had seen rents rise in the early 1870s and many gave up their farms as they could no longer make them pay.

Evans noted how Suffolk farmers handled the depression better than those in Norfolk, where particularly in west Norfolk there were many large farms, so the economic impact was far greater. Many farmers gave up, and owners of some of the big estates, often several thousands of acres, found it almost impossible to find tenant farmers willing to take the land on. The late Leonard Mason took over vast stretches of west Norfolk. His family still farm extensively in west Norfolk today.

Most of the people Evans interviewed were born in the last quarter of the nineteenth century. They will have been children during the Great Depression and so few spoke about it. An exception was John Goddard, who died in 1953 at the age of ninety-eight. Born a farmer's son, his father set him up with a sixty-acre holding in 1876, when he was twenty-two. 'Young Goddard had to pay £2 an acre rent, plus tithe charges. Stock and gear was also expensive: it was impossible for example to get a useful horse under £50.'[59] His father probably could not have foreseen the recession and his own farm slumped in value that year from £5,500 to less than £2,000. John found it impossible to make a living and took two part-time jobs, of 'Parish Overseer of the Poor and Surveyor of the Highways in his village.'[60] These paid him £20 a year, and together with some extra land his landlord let him take over without paying rent, he was able to continue.

This experience of farming through the Depression made John Goddard an early adopter of new technology, keen to try anything that could make his farm more profitable. He bought one of the first Suffolk Punch steam-engine and plough sets, made by Garretts of Leiston, and named after the horse it was built to replace. To plough in this way meant having two steam engines, one each side of the field, with the plough being winched back and forth between them, and, with each engine costing £700 according to Evans, this cannot have been much cheaper than using horses, as you needed at least three men, one on each engine and one to sit on the plough and steer it. But steam engines were more powerful than a pair of horses, could plough the land deeper and cover twenty acres a day, compared with the horses, which could plough just an acre. Goddard kept using horses, as well as steam, until Fordson tractors began to appear in the 1930s.

These tough times saw many Scottish farmers moving down to East Anglia. I can remember from the 1980s, when I was selling fertiliser in Norfolk, that many farmers there had Scottish family names. Their forebears had moved down in the 1880s, or during the later depression in the 1920s. Scottish farmers in East Anglia were described as 'a sturdy race of farmers, unspoilt by prosperity, thrifty and hard working.'[61]

Moving lock, stock and barrel from Scotland to East Anglia was not simple; many migrating farmers hired a train to transport livestock and equipment. Travel today is taken for granted, with a motorway network and even the smallest rural road hard-surfaced. But in the 1880s the railway was the only way to travel any distance. Trains had largely replaced the horse and carriage and dramatically reduced journey times. Cambridge, for example, could be reached from London in an hour by train in 1880; before the track was laid in 1844, the stage-coach took seven hours, with many changes of horses, to make the same journey. Trains then must have seemed as innovative as sending an email is today, both of them enabling faster communication.

Farm workers enjoyed greater mobility than those who employed them. The working-class writer Fred Kitchen, who was born twenty years before Evans, wrote about attending the St Martin's Day hiring fairs, at which farmers and farm workers would negotiate who would work where and for what wage over the next twelve months. Once hired, a man was bound by contract to work for the agreed rate for a full year. Similarly, the farmer was bound to keep the worker employed until the following November. Kitchen worked on farms from his thirteenth birthday.

Kitchen lived and worked in South Yorkshire, where most farms were mixed, with both arable and grassland, so there was livestock as well as crops to tend. This meant that the workload varied less throughout the year than in Suffolk, where many farms were purely arable. A farm that grows crops, but has no livestock, will have needed many hands at harvest time, but apart from threshing, which was often done by a contractor, there was little work to do in the winter months. Suffolk farmers needed more flexibility; some workers, such as horsemen and stockmen, would work on the same farm for generations, and others moved around according to demand for their labour.

In the early days of the twentieth century, cereal prices began to recover, although by then 80 per cent of the wheat used in the UK was imported from North America. Raymond Keer, a farmer from Bealings, near Woodbridge, told Evans: 'A frequently quoted mnemonic saying among old school farmers was: A man's wages is equal to the price of a sack (or comb) of corn.'[62] It sounds harsh today to measure your worker's value in this way, but attitudes were different in those days. Keer described how a general labourer, paid twelve shillings a week, was the value of a twelve-stone sack of oats. A horseman, paid fourteen shillings a week, equated to a sixteen-stone sack of barley, and a head horseman, sixteen shillings a week, which was the value of an eighteen-stone sack of wheat. These were the weights of each grain a standard comb sack would hold.

The experience of John Goddard, born in 1855 at Tunstall, was that farm workers would travel to Suffolk to be hired for the harvest. His grandson Geoffrey told Evans how men would come down to Suffolk from Lincolnshire to take the harvest. 'These workers would get paid their earnest or hiring money of a shilling in the normal way after the contract had been signed, and they'd immediately go out and drink it.'[63] A shilling went a long way in the 1890s; it would buy up to eight pints of beer or a half-bottle of gin. Typically, a worker was contracted for either one month or on a fixed fee until the harvest had been safely gathered.

It was also not uncommon for men to follow the harvest, travelling down to Essex where corn was usually harvested before that in Suffolk. Rendham farmer Harold Goddard, speaking about this pre-First World War practice, rather unkindly said: 'Essex farmers harvested early as they were not so particular about the quality of their corn.'[64] But as is often still true today, the closer you get to London, the better the money. 'The worker in East Anglia during the Industrial Revolution and up to the present day sold his labour in a buyer's market: the nearer he was to London – up Essex way – the more likely he was to get higher wages.'[65]

Men with agricultural skills could also find work in London. Norfolk farmer James Seeley, reminiscing with Evans about people he had known over the years, explained of one: 'He used to make hay in some of the London parks. Before they had machines they used to cut the grass with scythes. At that time when there were no other transport than horses, they could use all the hay they could get up in London.'[66] Horses were used for transport in every city, so hay was grown as a cash crop as well as for use on the farm, particularly on farms within horse and wagon range of a city.

Albert Love, born in 1886, worked on a farm at Alburgh, sixteen miles south of Norwich: 'I used to take wagons into Norwich with hay and straw about 1908 to 1910. All the farmers used to sell so much hay and so much straw to different places in Norwich for their horses.'[67] Just as London today imposes a congestion charge to reduce traffic, so too did Norwich, which had a system to keep the traffic running smoothly: 'Approaching Norwich we'd keep watching the time, because as you were drawing in to Norwich police used to be standing out there; and if you weren't there by 10 o'clock in the morning – they wouldn't let you in.'[68] To get to the city in time, Love would leave the farmyard at two-thirty in the morning, stopping halfway at half past five for breakfast at the Bird in Hand pub at Tasburgh.

Norwich cattle market, which from 1738 until 1960 was held in the city centre next to the castle, was another cause of road congestion. James Moore was born in Norwich in 1896 and employed as a cattle drover for most of his working life. He worked for an Irish cattle dealer, collecting cattle that arrived by train on the outskirts of the city: 'About thirty to thirty-five wagons of cattle would arrive at Trowse on a Thursday. We would meet them and feed them until the following Saturday. The cattle came from all over Ireland: Roscommon, Waterford, Derry, County Meath and Clonmel.'[69]

Irish cattle were well suited to grazing the coastal marshes a few miles to the east of Norwich. They would come by sea to Holyhead, and then be carried by train to Norwich, arriving on a Thursday for that Saturday's market. They would spend a couple of days grazing on the Trowse riverside marshes before being driven through the streets to market. The rail journey from Holyhead took fourteen hours, so I suspect they benefited from being given time to recover from their journey. James Moore was enthusiastic about the Irish cattle, explaining that after just three months on the east Norfolk marshes, they had fattened up sufficiently to return to market, this time to be sold to a butcher.

When harvest was over, many men from Suffolk farms would travel to work in the maltings at Burton-on-Trent: 'The migration was a sizeable one. It was in existence as early as 1880. By 1890 Bass and Company alone were employing over 125 East Anglian workers. By 1896 it had risen to 256 workers.'[70] Some were persuaded to stay on the farm; James Knight, born in 1880, was a ploughman at Burgh, a village near Woodbridge. When he announced that he was leaving to go to Burton, the farmer said: '"You've got a good pair of horses, and you're a good man and I'll give you a shilling a week more than I give anyone else [. . .] I don't see what you want to leave for."'[71] Despite this offer, Knight left anyway. He was nineteen and had already worked for a spell on a farm in Yorkshire, before returning to Suffolk, so was keen to travel and see more of the world.

There were, wrote Evans, three good reasons why so many

men went from East Suffolk to work at Burton-upon-Trent. The 1880s was a time of recession, with many young men out of work; the Burton breweries were already buying barley locally and the malting season neatly filled the winter period, when fewer men were needed on arable farms. As John Kettle, who was born in 1894, of Framsden explained, 'there were about fifty of us young chaps in that area without work after harvest. Farming was in a bad way. You couldn't get work only at haysel and harvest.'[72]

Each August the brewers would write to those who had worked in Burton the previous year, telling them the days that the brewer's agent would be visiting the area to sign men up for the coming season. The agent would stay at the Station Hotel in Ipswich, or the Crown at Framingham. Once he'd signed his contract, a man would also be given a train ticket for the journey to Burton. For those making their first trip to Burton, this was usually their first time on a train. Sam Friend said how strange he had found it, having travelled the five miles from Cretingham to Framlingham station by pony and trap to catch the small train that ran back and forth the six miles to Wickham Market, to then find himself on a large train with compartments.

A few years later everything changed. The First World War saw thousands of young men leave the farms to join up, and half a million horses were requisitioned by the army. At the outbreak of war, the UK was importing 60 per cent of its food supplies, particularly wheat from North America and sugar from Germany, where sugar beet was being grown years before it was introduced in Britain. Cereal prices increased dramatically and more than 2 million acres of land were brought back into production. The Great Depression had seen much land left fallow, as it had become uneconomic to farm it. German submarines sank more than 1 million tonnes of commercial shipping between October 1916 and January 1917, which further boosted the demand for home-grown produce.

James Seeley, born in 1894, joined up in 1914 and served throughout the war with the Eighth Norfolk regiment. He went to France in May 1915 and did not return to Suffolk until February 1919. 'We had proper training at the beginning of the war,' he told Evans, 'but I remember later drafts coming out to us – young lads from the Lancashire cotton mills. They joined the army, were sent to France and were killed, all in eight weeks.'[73] But Seeley described his days at the Western Front – in spite of the squalor and horror – as among the happiest of his life. He gained the rank of sergeant and was awarded the Military Medal for bravery in combat.

When he returned to the farm, his wages were higher as farm prices for wheat and oats were guaranteed by the 1917 Corn Production Act, which protected farm incomes until after the 1922 harvest. However, it was repealed a year earlier, in 1921, as world trade returned to pre-war levels, much to the dismay of many farmers. Cheap wheat from North America once more flooded the UK market, causing the price of home-grown wheat to halve. The men who came back from France saw their old pre-war lives through fresh eyes. 'The war had changed the men who had been in the army. They were better educated. They were not going to do the same things or put up with as much as they'd done afore they went out,'[74] Seeley told Evans.

James Seeley continued working on the farm until 1937, when he took the tenancy of a seventy-acre farm nearby. The land he took on had been neglected, with overgrown hedgerows and full dykes. Farming was in the doldrums throughout the interwar years, so for the brave it was a good time to set up on your own. Unemployment was high, and Seeley talked of a stone pit behind his farm where unemployed men were put to work by the county council. Married men could work there for three days a week, and unmarried men for just two.

In 1939, when Britain was once again at war with Germany, the War Agricultural Committees were reformed. They had the power to increase by 1.5 million acres the land available for food production. Each county had a target acreage of land to be cropped, and once more cereal prices increased, many men chose to follow their friends, leave the farms and join up, even though farm work was a reserved occupation, so exempt from conscription. Strangely unmentioned in any of Evans's books was the impact on Blaxhall of the arrival of the American air force. Bentwaters airfield, just two and half miles from the village, became operational in December 1944 and remained active until 1993, when it closed. The comedian Tommy Trinder, who was popular at the time, described the American servicemen as 'overpaid, overdressed and over here'. They must have been hard to ignore.

Farming in East Anglia continues to evolve, as I discovered when I visited Berry Farm, deep in the area of Suffolk known as The Saints. I'd seen Hayley Chitty and the farm featured in *Folk Features*, a weekly e-zine published by freelance journalist Emma Outten. Berry Farm is a cooperative, with, as I'd read, a campsite and shepherd's hut holiday let. Was this another example of the return of people to the land, as I'd seen at Flixton and Woodton, where farm diversification into cheese-making and a café had created new jobs? I decided to visit and find out.

The Saints is a group of small villages, all named after saints, that sit in the triangle between Beccles, Bungay and Halesworth. It's an easy place to get lost, with a tangle of twisting narrow lanes. On one of them I met a car coming the other way, and we sat for a moment facing each other down like gladiators, seeing who would reverse. I did the gentlemanly thing.

Berry Farm is just twenty acres, along a particularly long, meandering lane, with few houses and farms along its length. Fortunately it has a roadside sign, important for visiting campers, who otherwise might never find it. As I parked in front of the barn I could see cattle grazing behind the polytunnels and trees on the horizon.

Hayley made me a cup of tea and told me how a group of environmentally minded friends had formed two cooperatives, one to buy the land and the other to grow and market produce. The group hoped that they would eventually be able to live together on the farm, as several lived some way away. However, so far, only one home has been built, which took some effort as planners needed convincing that someone needed to live there to care for the animals and manage the cropping. Hayley's home was modest in size, timber-clad, and I suspect a lovely place to live, in a managed, but certainly not manicured, grassy corner of the farm.

A pond had been dug to collect water for irrigation, and a solar-powered pump lifted it into a tall tank from which a network of pipes ran to water the vegetables growing beside the polytunnels. With little passing traffic, selling produce at the farm gate would never work, so Berry Farm sell veg boxes that can be bought online and delivered to the door; most regular customers live within ten miles of the farm. The four Red Poll bullocks will one day feature in the Berry Farm meat boxes that are usually purchased by the veg box customers.

Knowing that ventures of this kind can be at the idealistic end of the organic farming spectrum, it was reassuring to learn that, while Berry Farm does not use pesticides or fertiliser, they've not felt the need for organic certification. Similarly, while some members of the cooperative are vegan, all are pragmatic enough to recognise the importance of keeping livestock. The cattle here lead a good life, and I think there's a world of difference between being grass-fed and reared on a smallholding, and enduring life on an American feedlot where cattle rarely get to see, let alone eat, a blade of grass.

What Berry Farm illustrates well is an approach to land management that would be familiar to the Victorian farmer, who used livestock and rotated his crops to maintain soil fertility. He managed perfectly well without fertilisers and sprays. The group running Berry Farm is also working in harmony with their land, respecting its biodiversity and using technology wisely. What had once been a twenty-acre wheat field, visited occasionally by a man driving a large tractor, was now a diverse enterprise, with many people involved in its cultivation. As I was discovering on my travels round Suffolk, this was no longer unusual.

POWER

Over the centuries we have become increasingly able to harness the world's natural resources. Coal has been mined since Roman times, and long before that, people learned to fashion iron into tools. Over the past 150 years, we have quickly moved from horse- to steam-power, on to oil, and now we are facing the dawn of an electric age. Mechanisation and automation have continued to take away the drudgery of the manual jobs that until the 1850s had changed little for centuries.

When George Ewart Evans wrote his books about life in rural Suffolk, the old ways of doing things were slipping from living memory. Those he interviewed had largely been born in the last quarter of the nineteenth century. Then, most people worked with their hands and the English class system made advancement difficult. Only in my lifetime has it become easier to get an education and follow a very different career from one's parents.

Blaxhall shepherd Robert Savage quite expected his sons to follow him and work with sheep, and indeed one of them, Russell, did just that, tending the sheep at on the Blackheath Estate where I worked in my mid-teens. But his descendants living in this century have led very different lives from those of Robert and Priscilla Savage. One I spoke with was an engineer, who before he retired employed a hundred men. Another has a senior role in the public sector, and a third a doctorate and a life in academia.

That pattern is repeated in my own family, with my maternal grandmother a housemaid in a country house in North Norfolk, where her father had been a woodman and her grandfather a blacksmith, all on the same estate. When in 1910 my grandmother was presented with a copy of *The Twins That Did Not Pair* as a school prize, she could not have imagined that one day a future grandson would be a published author.

But this social mobility has taken us further from the land and the seasons, and we have lost our connection with the natural world around us. Materially we are far better off than our forebears, but I'm not sure we are as happy as people once were. Can we turn the clock back to rediscover that sense of belonging that has today largely been lost?

Coal

I was six when we moved to Needham Market. The bungalow my parents bought was brand new; so new that I can remember standing in what was to be my bedroom before the roof went on, and there were just window-height walls to show the layout. As was usual in the early 1960s, the place did not have central heating; instead there was an open coal fire in the living room and a single-bar electric heater fitted in the bathroom, above the door. I suffered from chilblains every winter and can clearly remember ice forming on the inside of my bedroom window. This was normal.

People have been using coal to keep warm for 2,000 years. Sea coal, washed onto the Tyneside beaches from coal seams on the sea bed, and coal dug from open pits, were used by the Romans to heat the public baths that were as much social centres as places to get clean. As Britain's forests were cleared, wood and charcoal became more costly and coal became the fuel of choice, although as long ago as 1306 legislation was passed to discourage coal-burning in London, as the smoke from open fires was already becoming a problem.

The Industrial Revolution saw a rapid rise in the consumption of coal, which became the obvious fuel for blast furnaces and the steam engines that powered the cotton mills in Yorkshire and Lancashire. Coal mining became a major employer in South Wales through the nineteenth century, with 250 mines dug between Llanelli on the west coast and Blaenavon, just twenty miles west of the River Severn and the English border.

One of those mines was developed in 1889 at Abercynon, a small village fifteen miles from Cardiff and 270 miles west of Blaxhall. Abercynon is where George Ewart Evans was born in 1909. He was the son of the local shopkeeper, although his father William had started out working in the mine at the age of nine until a foot injury – 'he had a club foot sustained as a brattice boy, opening and closing the heavy door (the brattice) for pit ponies at the colliery'[75] – meant he could no longer work underground. 'He became apprenticed to a grocer in Cardiff, before moving to Abercynon in the late 1880s to start his own business.'[76]

To me, raised in rural East Anglia, a mining village in South Wales was an alien world of hills, moorland and tight clusters of terraced cottages built to accommodate the miners. The landscape I grew up in is very different, with wide-open landscapes, intensively farmed land and nothing a Welshman would ever consider a real hill.

I was thinking about these differences one day while at the dentist. I'd recently arranged a visit to Abercynon to see for myself the place where Evans grew up, studied and started writing. To distract me from the tooth that was about to be filled,

I'd taken along my copy of *Ask the Fellows Who Cut the Hay* to pass the time.

My reading was interrupted when the only other person in the waiting room, a woman of about my age, suddenly introduced herself, saying that the book I was reading had been her father's favourite. It was, she told me, written by someone who had grown up in the same village that he had. Startled, I asked if she meant Abercynon and she said that her father had indeed been born into a family of miners in that very place. We talked and later met for coffee, finding that we had several common interests in Norfolk, as well as that connection with George Ewart Evans.

Her father had not moved to East Anglia as Evans had done, but her husband's career had brought them to Norfolk, where they were now living. Today it is not at all unusual for people to move to other parts of the country as their careers develop, but when Evans was a boy, and a few decades earlier, when those he later interviewed in Suffolk were young, few moved far from the place where they were born. Some Suffolk farm workers might go to work in the maltings at Burton-upon-Trent in the winter, but they would return home when work on the land became once more available the following spring.

Just as my visit to the dentist had delivered a surprise, so too did my journey to Abercynon to see where Evans had grown up and to meet his biographer, Gareth Williams, an emeritus professor at the University of South Wales. I went by train, which meant changing at Cambridge, taking the Tube from King's Cross to Paddington, and then a train to Cardiff. From there I took a small, rather old two-carriage train up the valley from Cardiff to Abercynon. The connections were all on time, so the journey was actually faster by train than by car.

The train that runs to Abercynon, and then on to either Aberdare or Merthyr Tydfil, as the line divides just north of the station, takes just half an hour and follows the bank of the River

Taff for most of the way. The train rattles and bangs as it travels slowly up the valley, with the trees on the hillside almost touching the train as it passes. At every turn there is a new view of the river below, fast-flowing and rocky. It was very different from the sleek, fast Hitachi express train that had taken me from London to Cardiff.

It rather felt that the train, which I later learned dated from 1985, was taking me back in time. George Ewart Evans had made the same journey twice a day when a student at Cardiff University in the late 1920s and even earlier, up to the next station on from Abercynon to attend Mountain Ash county school. The trains he took were powered by steam, not diesel, but the views from the windows will have changed little since his day.

It was on the railway that Evans had what I suspect was a significant introduction to trade unionism and the battle for workers' rights that was to be reflected in his future writing. 'On Tuesday 4 May 1926,' he wrote, 'the General Strike began, as was expected, and the same train taking us to school did not run.'[77] The strike lasted just nine days. Evans and his friend walked the four miles to school each morning, then back again in the afternoon.

The General Strike did little to improve the pay of the Welsh miners, and his father's shop saw turnover fall as a result. This and the opening of a new Co-operative store in the same street forced his father into bankruptcy. This would have been devastating for the family had his father not been friendly with Evan Jones, the Co-op's general manager. 'He persuaded his committee to give Father a job,' Evans wrote, 'and also allow him to live on in the house.'[78] This was the silver lining to the cloud that could so easily have swamped the family. Evans described this very human act as 'a lifeline and it was as though out of a sky of hundred percent cloud there had appeared a cleft of light instantly followed by a brief flash of sun.'[79]

This dramatic change in the family's fortunes contributed to Evans's choice of career. 'I found out that if I agreed to become a teacher I could get a grant to take me through Higher Schools and on to university,'[80] he explained, describing teaching as 'the only way I could get to university at all'.[81] I wondered what he would have studied had he had more choice, and if a different career path would still have led him to Blaxhall and success as an oral historian. But he met his wife when they were both teaching at the same school in Cambridgeshire, and it was her career that took them to Blaxhall. It's strange how our early decisions shape our future life. As a parallel, I might never have met my wife had I not wanted to keep hens as a boy.

Abercynon today looks very much as it must have done when Evans was a child, with terraced streets and several chapels, including the one where his father was a deacon. The mine closed in 1988, and as a consequence many people in the village now commute to work in Cardiff. The village has a population of around 6,000, and with a grant from the Welsh government, a 320-space car park was built by the station, making it easier for people living in the wider area to park and ride into Cardiff. The buildings at the former pithead have been converted into business units.

The shop on Glancynon Terrace where Evans was born is still a shop today. Now a Premier Stores, it proudly bears a blue plaque in recognition of the fame of its former resident. It is not a large shop, and I had a pleasant conversation with the couple who now run it. The counter and till are by the door, where I suspect they have always been. Evans was one of eleven children, so the modest house behind the shop must have been bursting at the seams.

There is a footpath that runs to the right of the shop, from where I could see the buildings behind the shop that, when Evans lived there, had housed both the horse used to make deliveries and Daisy the cow. 'Outside were the stables, the big warehouse and the cart shed. There were two stalls in the stable: one for the horse and one for the cow.'[82] I paused to reflect on the story of how Daisy had been much loved as a house cow, but had met an untimely death because of her wandering nature.

Another connection with Evans's youth was that I stayed overnight at Llechwyn Hall Hotel, which had been a farm since the early eighteenth century, occupied by the Thomas family. It was one of several hilltop farms visited by Evans and his brother, making deliveries with their father's horse and cart. The road rising from the village loops up the steep hill, then on to Llechwyn Hall and the farms beyond. In his autobiography, he describes how, when delivering to Lechwyn with his brother in a pony and cart, he would jump down to the road and take a short cut up the hill while his brother steered the pony round the loops as the road climbed the steep hill. At the top, he would climb back onto the cart and continue with his brother to deliver groceries to the farms along the lane.

My room was in what had once been the Thomases' coach house, and before dinner I enjoyed a beer on a terrace that overlooked the village. Once more I was struck by the fact that this was a view that would have been familiar to Evans, and perhaps one that he remembered when living on the flat land of Suffolk and feeling nostalgic. Over breakfast the next morning I spoke with a fellow guest, who had returned to Abercynon for a family funeral. Her family had lived in the village for generations, but now the last remaining member had died. She thought this would be her final visit as now she had no reason to return.

It was a tremendous help to what seemed to be fast becoming a pilgrimage to the birthplace of a writer I'd admired for almost half a century to spend time with his biographer Gareth Williams and local historian David Maddox. They were excellent company and knowledgeable hosts, having collaborated to produce a commemorative book, *Ask the Fellows Who Cut the Coal*, which was locally published in 2017. They took me for a walk round the village, pointing out the school Evans had attended, the chapel, the pithead and the remains of the lock ladder that was used by the barges taking iron from the blast furnaces at Merthyr Tydfil down to Cardiff, along the Glamorganshire Canal.

Evans missed the Welsh landscape when he moved to Cambridgeshire, and then Suffolk. In particular he enjoyed walking, where he said he felt freer than he did in East Anglia, moved by what he described as the austere beauty of the Welsh hills. When the family moved to Blaxhall and Evans was no longer working, because of his deafness, he would take long walks through the gorse and bracken of Blaxhall Common, perhaps venturing down to the Forestry Commission's plantations in Tunstall, the next village.

With his wife busy running the school and with four children, Evans had plenty to do. The village did not have electricity, and as well as 'minding the children; occasional help in the school, the perpetual round of filling oil lamps and their servicing, the drawing of water from the well,'[83] he had plenty to distract him from writing. This did little for his self-esteem; as he said, 'I was anxious to prove myself in my new capacity as a freelance writer.'[84] Evans was far from alone in finding procrastination too easy an alternative to actually writing. I certainly find it all too tempting to check my social media feeds, which are just a click or two away from the screen on which I compose my own writing.

But it was coal, not Twitter or Facebook, that became Evans's preoccupation when he first arrived in Blaxhall with his wife and young family in 1947. It was a particularly harsh winter and there was a national coal shortage. 'At one stage we had no coal except a heap of dust, and we were forced to improvise,'[85] he later wrote in *The Strength of the Hills*. He described how he bought some cement and welded the small coal into balls, which then went into the stove in an effort to keep the house warm. He'd seen people doing something similar in Pembrokeshire, but with clay not cement, during the General Strike of 1926.

Coal was the only reliable source of heat and national stocks were at an all-time low. It did not help that people in towns and cities were buying electric fires, which kept them warm but put additional demand on power stations, most of which were coal-fired. Eastern England saw deep snow and temperatures as low as minus-twenty degrees. It must have felt strange for a man raised in a mining village to be faced with a shortage of something he'd always taken for granted.

Snape Maltings, now a concert hall, is just two miles from Blaxhall, at the point beyond which the River Alde becomes unnavigable. Next to the maltings is the quay from which Osborne and Fennell ran a corn and coal business from around 1800. Tiny by today's standards, and tidal so only navigable at high water, the quay at Snape became a busy port taking cargoes to and from London. This was how coal first arrived from the North East and found its way to Blaxhall. One of Evans's early interviewees, George Messenger, who had been born in 1877, had worked as a barge man at Snape Maltings. In *Spoken History*, Evans wrote about how Messenger had loaded ships at the Snape quayside with malt, which was then taken round to the Thames and up to the London breweries.

Newson Garrett, grandson of Richard Garrett, who made steam engines at nearby Leiston, bought the warehouse and quayside in 1841. Collier ships had been bringing coal here for years, and locally grown grain had been shipped from Snape to Ipswich and beyond. Garrett also bought the Bow brewery in London, and a few years later in 1854 built the maltings at Snape to supply it.

The Garretts were an enterprising family. The tie rods that

stopped the weight of the barley pushing out the malting walls were cast at the Leiston works, and the bricks made at the family's own brickworks. Business was good and five years later a railway spur line was laid from Campsea Ashe to Snape, so that coal could arrive and malt leave by rail, although barges also continued to take malt to London until the Second World War. George Messenger would have spent time unloading coal and loading malt onto trucks. The railway line ran right into the maltings site, although Suffolk Punch horses were used to pull the wagons the last few yards. The railway engines would remain in the sidings just across the road, ready to make the return journey to the main line.

I met Clive Curtis, who had worked at the maltings when he left school. He told me that work started at six in the morning, with a break for breakfast at eight, then again for lunch at noon. He had lived in Snape so was able to go home for his meals. When the malting business went into liquidation in 1965, Clive became a painter and decorator. He still lives in Snape now opposite the village hall.

When Snape Maltings closed, the Gooderham family firm 'bought the Maltings complex, together with the Plough & Sail, 27 other dwellings and 32 acres of land'.[86] George Gooderham had been a miller in the 1880s, and when I was growing up Gooderham and Hayward had a small animal-feed business beside the A12 at Marlesford, five miles from the maltings and even closer to Blaxhall. My father-in-law was a customer and I suspect they supplied many of the farms in Blaxhall when they still kept livestock.

However, the Gooderhams could not make use of the 11,000-tonne malting kiln and that became the Snape Maltings Concert Hall, thanks to Benjamin Britten and Peter Pears, who had started the Aldeburgh Festival in 1948.[87] The festival soon outgrew the Jubilee Hall in the town and Britten saw the potential to convert the former kiln into a concert hall.

The Queen opened the new concert hall in 1967, but disaster struck two years later when it burned down. It was rebuilt and reopened in 1970. My wife Belinda had been a pupil at Aldringham primary school in 1967 and remembers the school taking everyone to the opening, to cheer and see the Queen. Later, when it reopened after the fire, her grandfather Bernard took her to the sound checks that required a full auditorium and what she recalls as a series of unmelodic noises. I'm sure there was an orchestra, but she remembers hearing loud cannons most of all.

Britten's music is also still heard in Aldeburgh. A moving performance of his opera *Peter Grimes* was staged on the beach in the summer of 2013. In the closing scene, where the disgraced Peter Grimes goes out in his boat to end his life, baritone Alan Oke left the stage in a small boat, sliding down into the crashing waves on the shore. The popularity of Aldeburgh and the surrounding countryside really started when the railway arrived in 1860, but it was Britten who made it the centre for the arts it is today.

Blaxhall, although very close to Snape Maltings, was untouched by its transformation from working maltings to concert hall. Those I've spoken with who were born in the village were pleased that the buildings, which had become quite derelict, were being used, but few talked of attending concerts there. There is, however, one strong connection between Blaxhall and the concert hall, with a clue provided by a brass plaque on the wall in the entrance foyer.

The brass plate was placed there in memory of Bob and Doris Ling, a couple who managed the concert hall for twenty years. Bob Ling was born in 1920 to one of the many Ling families living in Blaxhall. He started working in the maltings at the age of fifteen and was the fourth generation of his family to do so. His great-grandfather had been employed by Newson Garrett, so had worked there in the very early days.

Bob would spend days carrying sixteen-stone sacks of malt to load barges on the quayside, despite weighing just eleven stone himself. He was one of seventy men to lose their jobs when the maltings closed, and then became a gravedigger, working for himself and digging graves in local churchyards. Later he was invited to return to the maltings as Concert Hall manager, a job where no doubt his local knowledge and practical approach proved invaluable. He came to know Benjamin Britten and, touchingly, was invited to dig his grave at Aldeburgh when the composer died. He lined the grave with reeds taken from the riverside by Snape Maltings, because Britten had wanted to be buried at the maltings but this had not been allowed.

Blaxhall is surrounded by acres of heathland with few trees and many gorse bushes, so most families bought coal to heat their homes, having just one open fire in their sitting room. This was how things were in the early 1950s, when Evans was researching his books, and still the case a decade later when I moved to Needham Market. In the Blaxhall Evans knew, water for washing clothes or the weekly bath was heated in a large cast-iron cauldron called a copper. This was usually situated in an outhouse behind the cottage. The fire beneath the copper, and also the fire lit in the bread oven in preparation for baking, would have used too much precious coal, so whin faggots, bound bundles of dried gorse, were usually used.

I can remember my grandmother talking about her childhood home having very similar routines, although as her father was a woodman on the Gurney Estate at Northrepps, near Cromer, they did not need to gather gorse stalks to fuel the bread oven or copper as he could take home all the firewood he needed. Once, when I was taking my elderly uncle to see where his mother had been born, he leapt out of the car and pushed his way through the bushes opposite what had been his grandparents' cottage. I followed him and was shown the remains of the sty where, he told me, he used to like to talk to the family pig when staying during school holidays.

Childhood memories have a habit of staying with us throughout our lives, and we remember them more and more as we grow older. There is little an old person seems to enjoy more than the opportunity to talk about how life was when they were young. This was how Evans gathered the stories that so brought his books to life. Now, as I move through my sixties, I find myself looking back more and more each year.

I was keen to find people in Blaxhall who had known Evans, and struck gold when browsing the notices and books at the back of Blaxhall church. There was a copy of *Blaxhall's Living Past*, which had been compiled and published by the Blaxhall Archive Group in 2007. The book was illustrated with pictures of village life and gave a valuable insight into how things had continued to change after Evans moved away. The book's foreword had been written by the four Evans children, who of course had grown up in Blaxhall.

A note on the front of the book gave a phone number to ring for anyone interested in buying a copy. The price was £20, but I thought the book would be worth buying and would help me in my research. I called the number and it was answered by Daphne Gant, who told me that she lived just over the crossroads by the church. She had a copy I could buy if I wanted to pop over. Daphne was clearly pleased that someone had shown an interest in the village and its history, and so we talked for a while. To my delight, she explained that she was a granddaughter of Robert and Priscilla Savage, who feature prominently in *Ask the Fellows Who Cut the Hay*, and that she had lived in Blaxhall all her life. In fact, her other grandparents, whose surname was Ling, had lived in the cottage next door.

She remembered the Evans family, and David Gentleman, Evans's son-in-law who had illustrated most of his books, had once sat in Daphne's garden and painted a picture of the field behind. This connection encouraged me to agree a date for me to return to talk more with Daphne and her husband John, who had worked on one of the farms in the village.

Over tea and cake, Daphne told me how as a child she often used to help her grandmother Priscilla bake bread. 'People were self-sufficient in those days, and what you didn't grow or make yourself, you would get from your neighbours, giving them something they needed in return.' Daphne told me that she was sad that the village is no longer the close-knit community it once was. Today many of the houses in the village are second homes, occupied only at weekends and during the concert season at Snape Maltings.

You get a sense of just how tightly knit Blaxhall was in the 1950s when you read what Cyril Herring, who was postmaster in neighbouring Tunstall, told Evans: 'a strange postman delivering letters to Blaxhall would find it an absolute bedlam to deliver among the forty Lings and twenty Smiths'.[88] He went on to explain that 'very few people leave Blaxhall and most families are intermarried'.[89] The postman's challenge was made worse because none of the houses was numbered and few had a name. There were, for example, 'two J. W. Lings living in separate houses, both on the same Common'.[90]

Blaxhall was, and still is, a very scattered village and to deliver twice a day to all 120 houses would take five and a half hours. Priscilla Savage's maiden name had been Ling, as had Daphne Gant's mother's family. Father and son Toby and Kenny Ling were both tractor drivers on Blackheath Estate when I was working there, and I am sure were members of that same extended family. And of course the estate's shepherd Russell Savage was also from Blaxhall. Families are no longer as close-knit as they were in the 1950s.

Evans soon realised that this remote Suffolk village was similar in many respects to Abercynon. In both places, families had intermarried and few moved away. In Wales, sons followed their fathers down the mines, and in Suffolk most men worked on the same farms as their fathers. These connections, and the ancient words that had remained part of the local dialect for centuries, captured his imagination. From his early forties until his death in 1988, Evans dedicated his life to collecting and writing the recollections of people who lived and worked in rural East Suffolk.

But his interest in collecting what he called 'oral evidence' started shortly after he graduated from Cardiff University in 1930. The completion of his studies coincided with the global depression that had followed the stock market crash of 1929. There were no jobs for classicists, or for that matter anyone else. With no job, Evans took long walks in the hills above Abercynon. 'I had often met groups of miners going along the old Roman road that ran along the spine of the hill above my home. If there was a cool breeze I would sometimes find them squatting on the lee side of a dry stone wall. Here almost any topic under the sun was likely to be tossed about in the course of a morning's walk.'[91]

These old miners would talk about how mechanisation had transformed coal mining, with long conveyor belts replacing pit ponies and steel bracing replacing traditional timber pit props. More coal was now being extracted by fewer men, and where once every man could expect to find work in the mine, now there was no guarantee of work. Evans wrote about how even before they left school, boys would be employed on the surface, picking slag from the coal as it was moved on conveyor belts around the colliery. When they left school they would often work alongside their fathers, as each miner then had a boy working behind him, to clear away the coal as he cut it from the coalface.

It was the stories these miners told that fascinated him, and when, years later, Evans found himself again without a job when he moved to Blaxhall, he would once more start collecting the stories of those he met on his walks round the village. That led to the publication of *Ask the Fellows Who Cut the Hay* in 1956, a book that remains in print to this day, and a new career as an oral historian. Coal as a fuel has had its day and soon we will no longer be able to buy it to burn in our living-room fireplace. So much has changed over the last fifty years, but people have not really changed at all.

Steam

When I met Daphne and John Gant at their Blaxhall cottage, I had hardly sat down before I was offered a cup of tea. The kettle was already on when I arrived, as I suspect it is in most homes where visitors are expected. In many households a whistling kettle is synonymous with hospitality. It seems to be part of our nature to offer hospitality and inevitably, in Britain, that means tea!

I can remember that childhood visits to my grandparents also started with a pot of tea, although I usually had milk or orange squash. Only when approaching my teens was I considered old enough for hot drinks. I grew up seeing coffee and tea as adult drinks, although this quickly changed when I discovered alcohol.

Kettles were also symbolic of success for those Evans

interviewed. A popular pub pastime in the early twentieth century was playing skittles. Sam Friend described how they would always bowl for a copper kettle, preferring it to the cash alternative that was equivalent to a week's wages. In books, films and pictures of bygone times, there is always a kettle gently bubbling on the stove. Cézanne chose to include a kettle in a still life he painted in 1869 during his dark period, and the kettle has long been in idiomatic usage in the English language. We talk of kettles of fish, pots calling the kettle black and so on.

There is a very good reason why people used to boil water. Until relatively recently water was drawn from wells and often contaminated. Boiling water kills most of the pathogens that might be lurking there. It was Victorian physician John Snow who, when investigating a London cholera outbreak, traced it to a contaminated well in Broad Street. More than 500 had died in the space of two weeks, but when the handle was removed from the well, making it unusable, no further cases occurred. He is widely considered to be the founding father of epidemiology, the branch of science in which my daughter works.

We all know that when a kettle boils, a jet of steam emerges from the spout, but it was not until the closing years of the seventeenth century that its potential power was recognized. The engineer Thomas Savery was perhaps the first to patent a steam engine that was used to draw water out of coal mines. Previously water had been lifted away from the coal seam by a horse-powered system of buckets and pulleys. Coal from the mine heated the water which generated steam and powered his machine.

Years later James Watt further developed the steam engine, and through the Industrial Revolution steam power replaced wind and water as the power source of choice. Steam power gave us the ability to generate power where and when it was needed, which transformed mining and manufacturing, and later, in 1804, Richard Trevithick successfully mounted a steam engine on wheels, replacing the horses that had previously pulled wagons of iron along the tramway that ran from Merthyr Tydfil to the canal at Abercynon. When I visited the town I was shown the monument that commemorates this fact. It stands in front of the fire station, close to where the old tramway ended.

It struck me as a fascinating coincidence that this very first steam train had journeyed to the place where Evans was born. Blaxhall, where he later lived, is near Leiston, where Richard Garrett set up the Long Shop, the world's first production line. Here steam traction engines were built in ever-increasing numbers by a rapidly growing workforce. A very contemporary application of steam power was the nuclear power station at Sizewell, two miles from Leiston. Sizewell was not the first nuclear power station; that was Calder Hall at Windscale, which had opened nine years earlier in 1956, but it brought Leiston's relationship with steam right up to date.

To me, Leiston will always be Suffolk's steam town, first put on the map by Garretts, which, when I lived there in the late 1960s, was replaced by the power station as the town's largest employer. People there smile when I say this, but I believe that steam defines the town, just as Hay-on-Wye is known for books, or Aldeburgh, where Benjamin Britten lived and founded the music festival that led to the later transformation of Snape Maltings, is known for music.

There is a large sculpture known as *The Scallop* by Maggie Hambling on the beach at Aldeburgh, just in case any visitors forget the town's connection with the arts. Ironically there are no scallops in the sea off the Suffolk coast, and Britten I'm told was allergic to shellfish. But these practicalities did not interfere with Hambling's creativity and *The Scallop* remains an iconic, and not always popular, reminder that Aldeburgh today is far more than the quiet fishing town it once was.

When Evans was a boy, the Abercynon coal mine employed more than 2,500 men, so will have dominated life in the town. His father William had been born in 1866 at Pentyrch, which is now a suburb of Cardiff. As a boy he had worked in a nearby mine where the Rhondda coal seam ran under the hills to the north of the village where he lived. Coal had been dug here since the late eighteenth century, so it is inevitable that the winding gear that lowered men in and out of the ground will have been steam-powered. This innovation allowed mines to be dug deeper.

In fact, according to Abercynon historian Keith Jones, the winding gear at the town's colliery was steam-powered until 1957. He explained that it was not until the early 1960s, when the National Coal Board were investing in a large modernisation programme in South Wales, that the banks of steam boilers were replaced with electric motors. The steam boilers had to be manned night and day to be ready when needed. Electric motors, on the other hand, can provide instant power with the flick of a switch.

Demand for coal rocketed during the nineteenth century, and by 1850 production from the Welsh mines had increased by 500 per cent. It's hard to imagine now what it must have felt like to live in a time of full employment and, I'd like to say, increasing prosperity, but working in the coal mines was dirty and dangerous, and wages were far from generous. The mine owners grew wealthy, but the miners themselves just about survived.

Today, there are no working coal mines in Britain, although, controversially, a new one is planned near Whitehaven in Cumbria (the claim is that this will be greener than shipping coal thousands of miles from North America or Australia for the remaining British steelworks, but the hope is that even they will be using electric furnaces from 2035). Perhaps not on the government's radar are the heritage steam railways and vintage steam engines that bring history alive for so many children. All now have to burn costly imported coal, and blacksmiths have to import coke or use gas-fired or electric furnaces. The Cumbria mine will produce more than 2 million tonnes of coal a year, which sounds a lot until you realise that in the 1980s the UK produced 128 million tonnes of coal annually.

Visiting old school friend Dafydd Aubrey Thomas in the 1960s gave Evans the opportunity in later life to interview men who had worked underground. One, John Williams, had started work as a boy in 1918, working alongside his father: 'There was one disadvantage to working with your father: he was not giving you any pocket money out of the pay'.[92] Miners' wages were set by the Coal Mines (Minimum Wage) Act of 1912. Boys under the age of fifteen were paid just one shilling and sixpence a day, rising to three shillings a day at the age of twenty. Farm workers in Suffolk by comparison were earning around five shillings a day.

It is hard to imagine the optimism and hope that defined the nineteenth century. Steam power meant greater mechanisation, freeing manufacturers from any reliance on wind speed or water flow. The British Empire was growing, bringing affordable raw materials such as cotton, and creating new markets for British-manufactured goods all round the world. Slavery may have been abolished in the British Empire in 1839, but from today's perspective, much of Britain's prosperity was the product of the exploitation of other nations. But we cannot change the past, only the future.

The UK also saw a rapid growth in population, which tripled from 6 million in 1750 to almost 18 million in 1850. This was in part due to improving health care and reducing child mortality, but also due to large numbers arriving from Ireland in search of a better future. Although in Blaxhall most people still worked on the land, they cannot fail to have been touched by this Victorian expansion. Snape Maltings started operation in 1846 and Richard Garrett's Long Shop opened in 1852. While earlier it had been inevitable that sons would follow their fathers onto the land, perhaps even then young men had the opportunity to choose between farm and factory.

George Ewart Evans also saw his life transformed by new technology. As an oral historian he would have taken notes as those he was interviewing spoke about the past. He had a hearing problem, which must have made this particularly difficult, and of course the broad Suffolk dialect people spoke was very different from the way people spoke in Cambridgeshire where he had lived before moving to Blaxhall. His life must have become much easier when he acquired a portable tape recorder.

It was the BBC that gave Evans a Marconi L2 Midget portable recording unit. You can see the machine today if you visit the Museum of East Anglian Life at Stowmarket, where it sits next to his typewriter and hat in a glass case under his photograph in Abbot's Hall. As Evans wrote in 1987: 'The Midget had its own battery-power, which was very important 30 years ago as mains electricity was far from being completely supplied in all the villages.'[93]

The recorder made it possible for some of his recordings to be broadcast by the BBC. Today they form part of an extensive digital archive of Evans's work at the British Library. He did not at first see the opportunity to transcribe the tapes for his books. 'I recognized at length it was legitimate to translate the full flavour of the recordings to paper. By this time I had become familiar with the Suffolk dialect and had come to appreciate its special quality.'[94]

He would not have wanted to guess at words, so to transcribe accurately he first had to really understand the Suffolk dialect. When I was growing up a few miles from Blaxhall, I could tell the difference between the Ipswich accent and that of people living in rural East Suffolk. Even fifty years ago, people did not more far, so could be placed by the way they spoke.

With his recordings, he took part in a series of radio programmes that were broadcast from the BBC's Norwich studio. He found the title *Through East Anglian Eyes* a little ironic, perhaps considering that *Through East Anglian Ears* would have been more accurate. The programmes were broadcast on the BBC Home Service between March 1959 and November 1960 and featured a number of well-known East Anglian characters such as the Norfolk nurseryman Alan Bloom, as well as Evans.

A career in radio could have followed, but he found that as television was growing in popularity, sound broadcasting was changing and less work was being commissioned. The recommendation of American management consultants, advising the BBC on how to become more efficient, led to what was to have been an hour-long programme about the annual migration of Suffolk farm workers to the maltings in Burton-upon-Trent being cut by half; 'moreover,' Evans explained, 'it would be for only half the fee and therefore a poor return for the time and trouble that had gone into it.'[95] At the time, Evans was deriving a third of his income from the BBC, and wrote disparagingly about how this American influence meant that 'British television and sound broadcasting lost the undoubted cultural centrality and international reputation it once possessed.'[96]

Evans had served for a few years on the Radio Writers committee of the Society of Authors, and so was in touch with the growing anxiety this group felt about the way things were going, particularly those who relied on writing radio scripts to make a living. 'Television was now completely in the saddle,'[97] he wrote, 'and the Third Programme was effectively scrapped. Before long it would have only the occasional feature, and would be filled with music.' In summary, he wrote that 'the word was being demoted'.[98]

What would he have made of television broadcasting today, where, with so many channels available, both free to view and by subscription, there is once more room for programmes on almost any topic? The difference today is that, with the exception of the BBC, most programmes are commercially sponsored. Had Evans set out today to make programmes about rural life, he might well have started by selling the idea to a commercial sponsor. That would not have resonated well with his left-wing politics, which he did not shout about, but were apparent in some of his writing. He was a regular contributor to *Left Review*. A piece titled 'Red Coal' was published in February 1937 under the name George Geraint, which Evans used as a pseudonym because he was worried his writing might interfere with his job as a teacher.

Technological advance has always caused controversy. The increasing mechanisation through the nineteenth century saw many lose their jobs, particularly on the land and down the mines. New jobs were created in factories, causing towns and cities to grow as people left rural England in search of a better life. Most of our towns and cities have street after street of Victorian terraced houses, built to accommodate the growing urban workforce of the day.

The house I lived in for most of my teenage years, above and behind Barclays Bank, opposite the Garretts site in Leiston, was surrounded by the terraced homes built to accommodate the factory workforce. The town grew significantly as Garretts expanded during the nineteenth century, but never in the expansive way of, say, Bournville or Saltaire.

Perhaps the Garrett family was less public spirited than George Cadbury or Titus Salt, or perhaps local landowners were not inclined to sell the land necessary for a more open townscape. Richard Garrett himself lived on the town's northern edge, in a grand house that in 1921 became the home of A. S. Neil's Summerhill School. Neil's school was as innovative and groundbreaking as the town-centre layout was not. Or perhaps I am too familiar with Leiston, so view it differently from others. The Garretts did, however, establish the Leiston Works

Athletic Association in the 1920s, with a sports ground, just over the road from the works. The name changed when Garretts closed, and it is now the Leiston Town Athletic Association.

Since the closure of Garretts, Sizewell nuclear power station has been the largest local employer. Like Garretts, it is all about steam, which is heated by nuclear fission and drives huge turbines that generate electricity. Huge diesel generators are there as back-up, but Sizewell is as much about steam in the twentieth century as Garretts was about steam in the nineteenth century.

There is a railway siding on the edge of Leiston, a mile from the power station where, for many years, fifty-tonne lead-lined flasks containing uranium were delivered to the first power station, and spent fuel was taken away for processing. Another connection that I find interesting is that had Garretts not been in the town, the railway branch line from Saxmundham would probably never have been laid, so it would not have continued as it did, to Aldeburgh, transforming the town from a sleepy fishing community into the elegant holiday resort it remains to this day. The line from Saxmundham to Aldeburgh closed in September 1966, just six months after Sizewell's first nuclear power station opened, and the track remains in place until just past the crossing gates.

Steam has once more become something of a hot topic in Leiston as it appears that a new nuclear power station, Sizewell C, is to be constructed. It will be built by French company EDF and be a sister plant to the one currently under construction at Hinckley Point in Somerset. There are roadside signs on the approaches to Leiston, erected by protesters that warn of heavy traffic and disruption, and none that celebrate the fact that this will create new jobs and give Leiston a much-needed economic boost. Freight trains, each carrying 400 tonnes of materials, are planned, and park-and-ride facilities for construction workers will be set up several miles away on the A12 to mitigate some of the negative impact.

This I suspect is very different from when the first power station was built at Sizewell more than half a century ago, but people have long memories. Back then a new road was built so that construction traffic did not have to go through Leiston, and the road to the A12 at Yoxford was improved, but there will still have been many hundreds of lorries on the road and 2,000 construction workers, many of whom lived in camps close to the site. The new power station will take nine years to build and, at its peak, employ more than 5,000 construction workers. It's easy to see why people are a little anxious.

There is an even stronger connection between the work of George Ewart Evans and the new Sizewell C power station because the land upon which it will be built was purchased from the Rope family, who farmed both at Leiston and Blaxhall.

The first two nuclear power stations at Sizewell were built on land owned by the Ogilvie family since 1859, when Scottish civil engineer Glencairn Ogilvie had bought Sizewell House, now called Sizewell Hall, as a holiday home. He'd made his fortune from building railways around the world. The current owner, also Glencairn, had been my father-in-law's landlord, owning his farm at Knodishall. He also owned Crown Farm Leiston, where I worked as a student in 1977.

The Rope family's 1,000 acres of marsh and heathland ran up the coast to the north of the Ogilvy Estate. Talking with Arthur Rope, one of the current generation, I learned that the family had originally farmed at Cransford, a village between Saxmundham and Framlingham. In the late eighteenth century, Richard Rope had been lord of the manor, so the family had both wealth and status. George Rope, born in 1814, had not been the eldest son, so left the land to make his way in the world as a corn and coal merchant, running a small fleet of ships from Iken and Orford. Iken is just a mile downriver of Snape, where Newson Garrett, a man just two years older than George, had built

Snape Maltings and used his own ships to take malt round the coast to London. They will have known each other and perhaps been friends.

At the age of thirty-two George Rope married Ann Pope, who had inherited Grove Farm from her late father. It was not a large farm, having fewer than 200 acres, so one of their sons, Arthur, had taken the tenancy of Lower Abbey Farm at nearby Leiston, which he had been able to buy when the owner died. Arthur's son Geoffrey was a man I remember from my childhood, as he, along with his two unmarried sisters, had worshipped at Leiston, usually attending the early morning communion service at which in the late 1970s I had often been the altar server. I seem to remember they would arrive rather regally, in a grey Rover 90. Geoffrey would then have been in his seventies, although to me, a teenager, he seemed quite ancient.

I have many memories of that time as an altar server. I used to share the rota for early-morning services with an old chap whose name I think was Fred. One Sunday morning he had arrived as usual, changed into his cassock and surplice, lit the candles on the altar and, feeling a little unwell, sat down in one of the front pews. It was here that, a few minutes later, the vicar John Drew, a tall man with an imposing black beard, found poor Fred sitting there dead. Years later, both of my parents had their funerals in that church, so I always associate it with mortality.

Talking with Arthur Rope about his family and how they had come to own land at both Blaxhall and Leiston reminded me of my own youthful ambition in that direction. When I was an agricultural student in the mid-1970s it had quickly become obvious that the only way to become a farmer was to marry the daughter of a farmer who had no sons. A number of my contemporaries had this goal clearly in their sights, and so the few farmers' daughters at Writtle Agricultural College were never short of potential suitors.

George Rope had been born into a landed family, but had farmed at Blaxhall because he married a farmer's daughter who, I think, had no brothers. I also married a farmer's daughter with no brothers, but her farm belonged to the Ogilvie family and so was tenanted. Years after he had retired, my father-in-law told me that, had I wanted to take over the farm, that might have been possible. This was never mentioned at the time, when it could have been negotiated, but in hindsight I would not have made a good farmer. I have a tendency to worry endlessly about things I cannot influence; so much in farming depends on the weather, over which I would have no control.

But I am not envious of the way the Rope family, or my peers at college, were able to become farmers by marriage. I had the benefit of spending a lot of my time on my in-laws' farm from the age of fourteen until they retired. I also have been very happily married now for almost forty years and have led an interesting and varied career, where opportunities presented themselves and were enthusiastically grasped. Not only would the uncertainties of farming have stressed me, but I would have become bored with the discipline and timetable the farming year would have imposed on my life.

Iron

The blacksmiths Evans wrote about played a central role in village life. They shod horses, put iron tyres onto cart wheels, and made or repaired all kinds of useful things for farmers and householders alike. The blacksmith's shop was also always warm, and so not unnaturally it was a place where men would hang around, at least until the pub opened.

In 1889 the Worshipful Company of Farriers set up a register of those qualified to shoe horses, and over the next thirty or so years qualifications evolved that focused as much on horse welfare as on the ability to craft and fit horseshoes. In 1923 an even harder examination had to be passed before you could become a Fellow of the Worshipful Company. This meant that a man entering the trade would make a choice between training to become a farrier, working with horses, or a blacksmith, making things at his forge.

By coincidence the 1920s were also seeing tractors beginning to replace horses on farms, and with fewer working horses to shoe, the farrier become the equine specialist that he is today. I visited Robert Rush, a third-generation farrier based at Clare in South Suffolk. His father and grandfather, also called Robert, had built an enviable reputation over the decades and although Robert Rush is not the only farrier around, customer loyalty is strong.

Robert told me that typically a horse is reshod around every five weeks. I have my hair cut every five weeks too, and so could see that just as I've used the same barber for years, horse owners will naturally keep using the same farrier, unless something happens to spoil the relationship. Robert explained how he and his team work hard to go the extra mile and deliver a great service, blending traditional skills with the latest technology. I was fascinated, for example, to see plastic shoes that are glued rather than nailed, to protect hooves recovering from laminitis, the painful inflammation of the connective tissue in the horse's foot.

As we talked, one of Robert's apprentices, Tom, was making a horseshoe using a portable gas forge rather than the more traditional coke-fuelled forge at the back of the workshop. He was expertly heating then hammering the steel bar into shape. Along one wall was a rack containing manufactured horseshoes in a range of sizes. These are not quite 'ready to wear' as some adjustment is still required when they are fitted. Robert's van carries a gas forge and travelling anvil as today the farrier goes to the horse, rather than the horse to the farrier. Robert's largest customer, Redwings horse sanctuary, is nearly sixty miles away, so he spends a good bit of time on the road.

We talked about how horse owners can be very particular, often appearing to care more for their horse than their own appearance, which reminded me of when I was calling on farms for a living: the wealthiest farmers were often the scruffiest! Clare is just fifteen miles from Newmarket, but apart from a few studs,

Robert does little work with racehorses, although he does carry a stock of the lightweight aluminium shoes they prefer. I picked up an aluminium shoe with one hand and a steel one with the other; the difference in weight was significant.

Horseshoes have long been considered to symbolise good fortune, and feature prominently in the design of wedding paraphernalia. Perhaps this links back to the famous blacksmith's forge at Gretna Green, where as far back as 1754 eloping couples fled the formalities of banns and a church wedding for something more spontaneous and perhaps romantic. I suspect it goes back much further; St Dunstan is said to be at the root of the belief that nailing a horseshoe on your door prevented the devil from entering. He lived more than 1,000 years ago.

Belinda and I were married at Aldringham church in September 1982 by the now late Charles Cowley, who was notable for having lost his right hand in an industrial accident and wearing a false one. It protruded from his cassock as if he was somehow sharing his vestments with a shop window mannequin. I seem to remember it had a moveable thumb he could clamp tight to hold his prayer book; he could then wildly wave his left hand, to emphasise the point he was making. He drove a twenty-year-old black 'sit up and beg' Ford Popular and enjoyed a pint or two at the Parrot in Aldringham after Sunday-morning services. He was something of a character.

George Ewart Evans did not have a church wedding when he married fellow teacher Florence Knappett in 1938. Money was tight and the couple married at Cambridge Registry Office.[99] They rented a flat, choosing a place they could afford, and, as Evans wrote, 'We were very doubtful about the set up, but we were eager to get married and did not ask too many questions.'[100] Their first home was the upper floor of a terraced house in Trumpington. It turned out that their landlord was in fact renting the house himself and did not have permission to sublet. They moved to a bungalow half a mile away shortly after.

I'm sure that horseshoes did not feature in their wedding, although in the late 1940s when living at Blaxhall, Evans became very interested in horses, which were quickly being replaced on farms by tractors. After the success of *Ask the Fellows Who Cut the Hay*, his first book about rural life, he wrote *The Horse in the Furrow*, which was published in 1959 and told the stories of men who had worked with horses in and around Blaxhall. Coincidentally I read his books about Suffolk in the order in which he wrote them. Over the years, I bought copies of them all.

Horseshoes also played a part in Suffolk tradition as Charles Bugg, who was born in 1884 and lived at Barking, a village just outside Needham Market, told Evans: 'Shoeing the colt' was a kind of primitive initiation ceremony.[101] This would take place when a boy entered his first field at the start of his first harvest. 'The men took you and turn you up and drove a nail into the bottom of your shoe. They would go on driving the nail until you shouted beer.'[102] This, he explained, meant that the boy had then promised to buy all the men a beer when he was paid his harvest bonus. I'm sure this was in reality light-hearted horseplay, but I suspect most lads shouted beer almost before the hammer struck its first blow.

As a keen observer of human behaviour and custom, Evans will have appreciated that this rather childish tradition had roots that may have stretch back centuries. Secret passwords and initiation ceremonies always seem to me to be a very English thing. My father had been a Freemason and because I became curious after his death, so too for ten or so years was I. But once I'd learned the complicated ritual and had my year as Master of a Lodge, it all began to seem rather pointless and so I resigned. Ritual, be it a masonic ceremony, Catholic mass or marking your first harvest, all seem to be designed to remind everyone who belongs, of their place in the pecking order.

Even something more straightforward, such as how to

position a horseshoe, is not as easy as it first looks. To bring you luck, it must be fixed with the ends pointing upward, unless you are a farrier, in which case it should point downwards. This must have been very confusing for a young man learning to work with horses. He will have been told one thing on the farm, and will have seen quite the opposite when visiting the farrier. Blacksmiths 'occupied a privileged position in the countryside because they shared the old horsemen's apparently magical secrets'.[103] Blacksmiths, Evans wrote, were admitted to the horsemen's secrets.

There is something strangely masonic about the way the Society of Horsemen conducted its meetings and initiated new members. Evans suspected that although primarily Scottish, members of the society had come to work in Suffolk, bringing the practices and customs with them. Just as a masonic Lodge posts one of its number, called the Outer Guard, outside its door, so too is the horsemen's place of meeting 'carefully guarded by two of the brethren who stay on watch for "cowans and intruders"'.[104] Those same two words feature in masonic ritual, which makes me wonder if there's a connection. I looked up the word cowan, and saw that it literally means one who builds dry stone walls. Masonic ritual revolves around the building of a temple, so there is a link.

Death is also a subject explored by both. A young horseman, unless he 'submitted to the conditioning ordeal was led off into the bush to die,'[105] or at least, as with Freemasons, to simulate the act of being struck dead by one of his fellow brethren. There is never any harm in reminding someone of their mortality and vulnerability. In both, the candidate is entrusted with secrets and pieces of folklore that can help him navigate life. For the horseman, this particularly taught him how to treat his horse, and to respect his fellow horsemen.

Where the Society of Horsemen clearly differed from freemasonry was in who could, or could not, join. Farmers and their sons were not allowed to join, even if already skilled horsemen, and as is so often the case, nor were women or girls. As critics of freemasonry are always quick to point out, membership often includes those with influence, which inevitably leads to accusations of favouritism. The Society of Horsemen was for the workers, and freemasonry the privileged.

I'm sure the horsemen that Evans interviewed also competed with their fathers. It is perhaps natural for the younger man to want to do better and prove himself; until the twentieth century, most boys followed in their father's footsteps, often working alongside them as they learned their trade. This is the way of the world in societies that live free of the burden of civilisation. What I was brought up to regard as primitive societies are in fact highly structured, often very hierarchical and well organised, as everyone knew their role and place. This was also the way in timeless, rural Suffolk until the early twentieth century when war, new technologies and rapidly improving transport changed things.

Evans drew parallels between the initiation ceremonies of the Society of Horsemen and what he described as the 'puberty initiation ceremonies of the primitive where the young boys are inducted into the clan and where the relation of the initiate and his fellows to one another and to the totem animal, representing the clan in depth, was defined and ritualised.'[106] It is interesting that he chose to describe the relationship between horseman and his animal as totemic, because in many ways it was. A man and horse would work together for perhaps twenty years, and so come to know and trust each other in the same way as you or I might form an attachment to our dog. The difference that made the horseman's bond even stronger was that his horse was crucial to his ability to work and support his family.

The Suffolk Punch was the horse of choice in the villages that Evans wrote about. It was the product of selective breeding, thanks in part to Henry VIII, who legislated to 'increase the size of the English horse'.[107] The 1541 Statute stipulated 'that no person shall put in any forest, chase, moor or heath, common or waste *any stoned horse* above the age of two years, not being fifteen hands high.'[108] My italics: I'd not heard the word 'stoned' used like this before, as a way of differentiating stallions from geldings, although when helping castrate piglets in the late 1960s, Mike Newson, who wielded the scalpel and on whose farm I helped one school holiday, always talked about frying the freshly removed 'stones' for his lunch.

All the Suffolk Punch horses today, and arguably all those worked by the men Evans interviewed, were descended from one stallion that was born in 1768, called Crisp's Horse of Ufford. It is just six miles from Ufford to Blaxhall, so not far at all for Mr Crisp to take his stallion to cover the mares. Until the sixteenth century, oxen had been used to plough fields and pull the four-wheeled wagons that carried the harvest back to the farmstead. The transition from ox to horse power was helped by the development of lighter ploughs. Where oxen used brute strength, horses were nimbler but not quite as powerful.

The earliest ploughs had been rather rudimentary with little apparent thought for efficient operation. The word plough is derived from the Anglo-Saxon word plog, and the earliest were pulled by ropes tied to an ox's horns. Later, harness were developed to use the power of the animal's shoulders, which must have made them far more comfortable for the poor ox. It was a Scot, John Small, who in 1763 applied some mathematics and science to the design of the cast iron mouldboard that turned the soil over when pulled through the ground. Later, in 1785, Norwich Quaker Robert Ransome patented the ploughshare he had developed and, from 1804, manufactured them at his Ipswich works.

Being a Quaker myself, it was no surprise to learn that it was 'with the help of Gurney's the Quaker bankers that he started a foundry business'.[109] Both Robert Ransome and John Gurney would have worshipped at Norwich's Quaker Meeting House and so known each other. Early Quakers had formed a tightly knit community, and the way they were organised, with local Meetings, Area Meetings and the British Yearly Meeting that took place annually in London, meant there was plenty of opportunity for business networking. At the time when Ransome set up his business, Quakers were still barred from attending university, which was why so many became entrepreneurs. Gurney's bank was the obvious place to go for a loan.

Arthur Biddle was a direct descendant of Robert Ransome and told Evans about how a chance discovery led to a further innovation in plough design. 'An accident occurred through the bursting of a Furnace, which gave way and all the molten metal ran out and some of it ran on to some iron plates and cooled rapidly – of course much more quickly than that which ran into the sandy ground of the foundry.'[110] Ransome saw that the iron that had cooled more quickly was harder. He called this effect 'chilled' and developed the technique of making ploughshares so that the bottom edge was hard, and the upper surface softer and kept sharp by the friction of the soil through which it passed.

It was this chilled iron plough that built Ransome's national reputation through the early decades of the nineteenth century. This development coincided with the replacement of oxen with horses, which I suspect was because these new ploughs were lighter and more streamlined, so easier to pull. Ransome's continued to innovate, developing the Yorkshire Light plough, known as the YL, in response to customer feedback. It had iron beams and handles and was, according to Evans, 'still being manufactured and sold in great numbers over a hundred years after its introduction in 1843'.[111]

Another member of the family was equally innovative. 'James Edward Ransome introduced one of the first motor tractors in

England in the year 1903; and during the following year demonstrated the petrol motor-tractor with the first tractor plough on Rushmere Heath, near Ipswich.'[112] This was at a time when a more common sight on Suffolk farms was a pair of Garrett steam tractors pulling a plough back and forth across a field on a steel cable. It was not until after the end of the First World War that the tractor began to replace both horses and steam engines.

'George Garrard who had a six-horse farm at Gislingham bought his first tractor in 1927.'[113] A six-horse farm was one with 300 acres, because each horse could plough fifty acres between harvest and midwinter. 'When he bought the tractor he sold two of his horses, working his land with four horses and the tractor. Later, he bought another tractor – a standard Fordson and sold two more horses.'[114] This was the beginning of a trend that has seen the number of Suffolk Punch horses in England fall from more than 2 million in the late nineteenth century to just 200 today.

One day I visited the Suffolk Punch Trust at Hollesley, which is just ten miles from Blaxhall on a sandy peninsula between the rivers Alde and Deben, close to the sea. Horses had been worked here since 1759, and while the estate has changed hugely since then, horses have remained to this day, although now more for show than practical use. In 1886 the farm was sold to become the Colonial College and Training Farm, where young men were taught the skills they'd need to make their way in America, Australia and New Zealand. In 1938 the site became a borstal, and fruit and vegetables grown by the inmates supplied a number of other such institutions, as well as feeding the young men who grew them.

Evans too had been to Hollesley to see the Suffolk Punch horses. There is a display in the museum there that celebrates the role he played in raising the profile of this powerful link with Suffolk's past. He met John Bramley, born in 1930, who had worked at the borstal since 1953. In his day a hundred of the young men serving sentences there worked on the farm. The farm then ran to 1,200 acres, with ten acres of nurseries, pigs, cattle and a large flock of Suffolk sheep. A pack house meant that vegetables could be prepared before despatch to other prisons and there was even a milk-pasteurising plant.

Showing him round the farm, Bramley proudly told Evans that 'we supply a quarter of the prison establishments with vegetables daily, using our own transport. We have our own vegetable processing plant and a jam factory.'[115] Although the farm had once had forty Suffolk Punches, when Evans visited there were just '21 working horses'.[116] He stressed the therapeutic role of the horses: 'We attach one boy to one horse and the boy's character changes: he has now got responsibility, direct responsibility.'[117] He went on to explain how most of the boys came from big cities, and formed a strong bond with the horse in their charge.

But as is so often the case, prison policy changed and learning horsemanship, which had given many young men their first real sense of purpose, was made redundant. A fundraising campaign raised the money to buy the stud and 188 acres of land, which is now owned and managed by the Suffolk Punch Trust. I was shown round by a volunteer who explained to me why the Suffolk horse was so well suited to field work; the Suffolk stands a little shorter than, say, a Shire horse, but has large, strong shoulders and a narrow rear end, which made it ideal pulling a plough or wagon. The Shire is also strong, but built differently and more agile, so was more often used for road work.

Even on farms that had mostly switched from horse to tractor, there remained a few jobs that were best done by the horse. William Cobbold, of Battisford near Needham Market, who was born in 1870, told Evans: 'I used the horses for ploughing the headlands after the tractors had ploughed the stetches; and for finishing up as we called it.'[118] He did this because ploughing those final furrows around the edges of the field would have meant running the tractor wheels on top of the recently ploughed land. This, Cobbold told Evans, could form a hard pan, which is a compacted layer of soil that water and roots find hard to penetrate. Today tractor-mounted sub-soilers are used to break up pans, but in the nineteenth century it had to wait until the following winter's ploughing.

In the early 1970s, Michael my father-in-law was also mindful of the damage machinery can inflict on a field. He and I would pull the sugar beet up in the corners of his field by hand, to prevent the tractor that was pulling the harvester from running over them and worse, compacting the soil. I can well remember spending cold weekends pulling the beet up with a beet hook, a long knife with a hook on the end, then cutting off the crown and leaves, before throwing the roots into heaps. The best job, although hardly less strenuous, was to return with a Fordson Major tractor and trailer onto which we loaded the beet with a beet fork (wide with rounded tips on the tines so as not to damage the beet). There was a technique to sliding the fork beneath the beet without picking up too much soil.

But of all the jobs on an arable farm, ploughing was always my favourite. It must have been 1972 when Michael showed me how to plough and finally, after much checking of my work, would allow me to drive to outlying fields to plough unsupervised. There is something innately satisfying about being on your own turning the debris from the old crop into the ground, creating a clean, fresh surface into which the following crop would be sown. The tractor would be followed by screaming gulls, eager to harvest the freshly exposed worms.

The tractor I ploughed with, a Massey Ferguson 165, was tiny by today's standards, with just fifty-eight horsepower, which meant it could only pull a two-furrow reversible plough. To me this combination sits halfway between the single-furrow plough that horses would have pulled fifty years earlier and the twelve-furrow ploughs that today's giant tractors can take in their stride half a century later. My tractor had a cab, which was open to the rear. This was important because at the end of the field, when the tractor had lifted the plough out of the soil, I had to reach back and tug hard on a metal bar that released the mechanism that swung the plough the other way up, ready for my return down the field.

A horse-drawn plough was not reversible, so the ploughman would plough the land in sections, called stetches. These would be carefully measured and were usually the same width as the Smyth drill that would drill the following crop. The field would be marked out first, then the first furrow drawn, and the second right beside it, going the other way. He would then go round and round until he had ploughed a strip of land the width of the drill. The centre of each stetch would be a little higher than the edges, which made it easier for the man doing the drilling to keep his horse and drill straight. Any missed strips would stick out like a sore thumb and be the cause of much ribaldry in the village pub.

A twenty-first-century tractor pulling a huge plough is steered by GPS and turned over hydraulically. When I ploughed with the Massey Ferguson I'd run one set of wheels in the furrow I had just created, but even then there was some skill required to keep the furrows straight. Today's ploughman simply programmes his GPS system and listens to the radio while his tractor sets its own course across the field. The horse learned to pull a straight furrow, I had to work hard to achieve it, and today precision is taken for granted.

If the ability a modern tractor has to set its own course across a field seems magic, so too were some of the customs faithfully followed by the old horsemen that Evans interviewed. I'd heard of horse whisperers, who are able to train a horse by developing an understanding of equine psychology. But crucial to the methods employed by the old Suffolk horsemen was a frog or toad's bone, considered a talisman that gave the horsemen greater power over even the strongest-willed horse.

Evans considered the use of this talisman to be rooted in witchcraft and followed because of the importance of the horse to those living and working in rural England. He pointed out that 'although the practitioners of horse magic were very few and far between they had a disproportionate effect on the corpus of beliefs in their community'.[119]

Jack Leeder, born in 1901, told Evans how the highly prized frog's bone was obtained, which to me illustrated the link to ancient witchcraft. 'It's a running toad,' he explained, 'and you would put it in an ant hill and so on, and when you are getting it – this bone that goes upstream – the Devil is there with you.'[120] A running toad is another name for the natterjack toad, and the ant hill was clearly a way of reducing the dead toad to a skeleton. The bones would then be carefully placed in a stream of running water, and one bone, the one that had the special powers, would float upstream, away from the rest. The horseman would carry this bone and was said to benefit from the special powers it gave him.

I'm sure that when Evans was listening to these old chaps talking with some conviction about how the old horsemen they'd known had believed in this magic, it must at times have been hard to keep a straight face. But he respected these old country beliefs and humoured those who believed them. His reasoning, he wrote, 'relates to the potential enlightenment they hold for the historian,'[121] going on to explain, 'I have become convinced that peasant attitudes and beliefs did not change for centuries; and it is only the younger generations that have changed appreciably within this current century.'[122]

It would, however, be a mistake to assume that all young people are quick to dispense with tradition. Having established that farriers shoe horses, and blacksmiths do everything else we associate with a forge, I wanted to see how blacksmithing had changed and so visited twenty-five-year-old Paul Stoddart, who runs the Kingdom Forge at Brundish, based in a remote barn a mile or so out of the village. The farmyard was home to several businesses and I was almost distracted by the County Super Six, a beefed-up 1960s four-wheel-drive Ford tractor standing outside the farm machinery business next door to the forge.

Clad in a leather apron, bearded and with a broad smile, Paul looked as I imagined a blacksmith should look: strong, happy and outgoing. The building was large, with four forges along one side, and assorted powered machines and steel taking up the remaining space. His anvil stood by the door, and had, I was told, once belonged to another blacksmith and been passed on when the older man had died.

Paul explained that, as well as making decorative ironwork for customers, he also ran blacksmithing courses that enabled people to experience making an axe, or other useful items, themselves. Men and women of all ages, from fourteen upwards, pay £160 and, suitably clad, spend a day at the forge, learning something of the blacksmith's skill and leaving with something they themselves have made. In 2019, the year before the pandemic, around 500 people took advantage of the opportunity. There is something deeply elemental about working with iron, fire and water in a way that has changed little over the centuries.

But while blacksmithing is a very traditional skill, most of the steel used today comes from East Asia and is of variable quality. Paul told me how students sometimes thought they had done something wrong because the bar they were hammering into shape split, while in fact it was a flaw in the bar that had caused the problem. Similarly, the coke that fired the forge is now imported from Poland. While it's important to reduce greenhouse gas emissions, it is a shame that traditional skills carried out on a comparatively modest scale are in danger of being lost.

Paul learned his craft at the blacksmith's school at Hereford College of the Arts. It's the UK's leading centre for this traditional skill, and as Paul told me, conveniently opposite the Bulmer cider factory, which is perhaps why I should not have been surprised to learn that with his father-in-law, Paul, he had recently launched Smith & Brew, a cider made from the apples that grow in the orchard behind the forge. Cider-making is another Suffolk tradition: Aspall Cyder was owned by the Chevallier Guild family from 1718 until 2018, when it was taken over by the American brewing giant Molson Coors.

Evans was sixty when *Horse Power and Magic* was published, the book in which he suggested that the younger generation was unlikely to preserve traditions that had previously remained unchanged for centuries. I wonder if he had become sceptical with age. He was writing about the early years of the twentieth century, but what, one wonders, would he have made of the current one, which already has seen some spectacular advances, both in the home and on the farm.

As recently as 2000 global positioning became commercially available, making possible the satellite navigation feature in your car and on your phone. It's how tractors now steer themselves with previously unimagined accuracy. It also allows the dosage of fertiliser and sprays to be adjusted across a field. This has become increasingly important as the government has committed to a 16 per cent reduction of ammonia emissions by 2030, prompting the likely ban of urea fertiliser, which is a major source of ammonia release to the environment.

The way we communicate has changed too, with people more likely to communicate with their friends online than chat with them over the garden fence. Evans died in 1988, when computers were becoming more common, but the internet was still in its infancy. The ability to build friendships, business networks and even find romance online has taken away the barrier of geography. Today it is as easy to meet, see and talk with someone on the other side of the world as it is to speak with someone living on your street.

These fundamental changes to the way we communicate, and the ease with which we can communicate with anyone anywhere, has altered our perception of community. It is no coincidence that so many people living in Blaxhall in the 1950s had the same family name. Most people did not travel far, and so courted people from their own or perhaps the neighbouring village. Today, this seems a little unimaginative, but it did mean that people had a network of neighbours living nearby to whom they were related and could turn to in adversity. There is much to be said for living in, and feeling part of, a strong local community.

COMMUNITY

Community has come to mean much more than the neighbourhood within which you live, and where you feel you most belong. The internet has removed the physical boundaries and made it easy for us to communicate with like-minded people, wherever in the world they happen to live. Today politicians, journalists and campaigners talk of the BAME community, or the gay community, conveniently grouping together people with just one thing in common.

In early-twentieth-century rural Suffolk, community was taken for granted and villages were largely self-contained and self-sufficient. Most had a village school, perhaps with just one or two teachers, the church played a key role in local affairs, and local tradespeople provided almost everything one needed to live and work. Trips into town were special occasions and often involved walking or cycling, as few owned a car and public transport was almost non-existent.

Today the internet makes it easy to find and communicate with people all round the world, but my sense is that for a community to really matter, it has to be something you live with more intimately. We may have little in common with people who live next door, but we have to get along with them because of that proximity. For those Evans interviewed, neighbours were of crucial importance. Until 1948 there was no National Health Service and social care was very different from what it is today. Most also married locally and so lived in communities where mutual support was taken for granted.

For me, and I guess most people, starting school was my first independent experience of living in a community. It was an opportunity to learn how to get along with other people and find common ground. Church, or for me my Quaker Meeting, was another, although in so many ways, is part of our cultural past rather than our present or future. Perhaps the local corner or village shop, where it exists, is the one of the last remaining places where we can feel valued locally, because we know that if we don't use it, it will soon be gone.

School

GEORGE EWART EVANS'S WIFE Florence was the only teacher who made enough of an impression on me still to be remembered after almost sixty years. It was her calm, quiet, encouraging manner that made her so memorable. She was also very fair and never to my knowledge lost her temper; even when someone had been naughty, she would gently chastise them in a way that usually prompted an apology. I don't think she believed in using the cane. Mrs Evans seemed to have a knack for seeing the good in everyone. Years later I realised that this must have been a manifestation of her Quaker faith.

I failed my eleven-plus, I think because I made an effort not to pass it, preferring instead to spend my time in the examination hall doodling. My father had made it clear he expected me to pass, and so failing was my first successful act of rebellion. For a year, I went to Stowmarket secondary modern school where my lasting memory is of being quizzed one afternoon by the headteacher, who, having heard that the French mistress had taken me to her house one lunchtime a few days earlier, wanted to know if anything untoward had happened.

A school trip to Brittany was for me a teenage voyage of discovery. Carentec, where we stayed, is roughly the same distance from East Suffolk as Abercynon, where Evans grew up and first went to school. Rural Glamorgan and coastal Brittany are in many ways equally 'foreign'; both have hills and are rocky, while Suffolk is flat and open. Both are also Celtic with their own languages, Breton and Welsh, and in both places people have a proud sense of identity. This may be a rather tenuous comparison, but just as my eyes were opened on that first visit to France, so too must Evans have been surprised by what he encountered when he found himself living in rural Suffolk.

Just as one or two memories of my time at school stand out, so too did Evans write about his earliest school days: 'I recall being visited at home once by my first teacher – the babies' teacher at infant school. My mother brought her up to my bedroom but I heard her voice as she was coming up the stairs and pretended to be asleep to escape the embarrassment of her perfunctory kiss.'[123] He clearly did not want his teacher to witness this natural show of affection.

Abercynon when Evans was growing up, at the time of the First World War, had a population of around 10,000. According to the 2019 census the population had fallen to around 6,000. This I suspect is a reflection of its change from a busy colliery town to a commuter town, a forty minute train ride up the valley from Cardiff. Families are also much smaller now than in Evans's day; in his autobiography he compared his father's fecundity to that of Job, who had seven sons and three daughters. Evans had six brothers and four sisters, which was far from unusual in the early twentieth century.

It was a wish to continue studying that led to Evans's training to become a teacher. After the family business failed it became vital to secure a grant or bursary of some kind if he was to continue at school and go to university. 'For when the shop closed it seemed to be expected of me that I should now leave school and earn a living.'[124]

Teaching had not been his first ambition, as he wrote: 'What I should have liked to do (this came about when I was ten years of age) was to go about examining castles and ruins – anything to do with the past.'[125] This rather revealing admission of a childhood dream, together with the interest in language that his classics degree gave him, make it no surprise that at around the age of forty, when living at Blaxhall and too deaf to teach, he quickly built a respected reputation as an oral historian. As he wrote, 'I laid down the foundation of the work I have been doing ever since; getting history directly from people, and going about introducing the technique to schools, continuing-education classes, and groups in universities all over Britain.'[126]

Following our destiny, rather than living the lives others expect of us, is important to our health and wellbeing. In his biography of Evans, Gareth Williams wrote that: 'Throughout the 1940s Evans was subject to frequent bouts of depression,'[127] he recalled. 'I felt I should have gone berserk were it not for my writing.'[128] For many years I have also lived with depression. Like Evans, I too have found peace in writing.

Evans's history teacher had the foresight to encourage him not to lose sight of his real ambition. She 'asked me what I was going to read at university and what I intended doing afterwards'.[129] She warned him of the danger of getting too settled as a teacher, falling into what she called 'the Rut'. The years can quickly pass once you get into the routine of a job. He suspected his teacher, who had 'a fine soprano' voice, rather regretted not following a career in music.

As I've grown older, I too have come to realise the importance of doing the work that I really want to do. My father had wanted my brother and me to join a bank when we left school, as he and his brother had done. My brother did work for a time at a bank, but quickly saw that it was not for him and soon moved to spend a long and satisfying career as a cost accountant with a large civil engineering business.

I never once considered banking as a career, changing direction a number of times before becoming self-employed at the age of thirty-five. This did not make me wealthy, but went a long way towards allowing me to follow my innate curiosity and constantly evolving interests. Now at an age when many retire, I have set off in yet another direction, writing full time. Only time will tell if this proves to be a new career, or just a kinder way of saying old-age pensioner.

I started school in 1960 and never doubted that I would continue through school and on into further education. I didn't go to university, at least not until I was in my mid-sixties, but did go to agricultural college, only starting my first proper job at the age of twenty-three. For those Evans interviewed, even staying at school until the age of eleven was not a given. It was not until 1918 that the government raised the school leaving age to fourteen, and then to fifteen in 1947 and finally in 1972 to sixteen. Today, you can only leave school at sixteen if you are going to college or starting an apprenticeship

Blaxhall school, where Florence Evans was headmistress, was built in 1881, closed in 1972 and is now a popular youth hostel. The school was probably built as a result of the 1870 Education Act, which had set up a mechanism for schools to be paid for by public taxation. Until then, schools had entirely been funded by local philanthropists, or established with the support of the Church of England. Blaxhall is what Evans called an open village, with many farms, rather than a closed village where all the land belonged to one family.

John Tollemache could afford to build Helmingham school in 1853 because his ancestors had owned Helmingham Hall and thousands of acres of land around it since the early sixteenth century. Florence Evans was headmistress of the school there for a time, and it was inevitable that George Ewart Evans would research its history. Writing about Tollmache, he explained, 'He built the school at his own expense; and he – and later his family – contributed generously to its upkeep until the county council took it over.'[130] Blaxhall had no single wealthy landowner and so building a school had to wait until the government would fund it.

Today, faith schools are the subject of some controversy, with many believing that education should be unbiased by religion of any kind. But in early Victorian Britain, were it not for the National Society for Promoting Religious Education, founded in 1811, which funded both the building and running of many rural schools, children would have had no education at all. It's hard to imagine, in today's largely secular society, how such a movement could emerge today. But in Victorian Britain, most people went to church on a Sunday morning and were more likely to turn to the church than the government for support. The society built and ran 12,000 schools, and even today, one third of all English primary schools, 4,605 to be precise, are owned by the Church of England, although funded by the state.

National schools were built for the poor, but perhaps understandably, when a wealthy landowner funded a school, he would have believed that, just as he will have had a privileged upbringing, so too did some in his village deserve a better education than others. As Evans wryly noted: 'there were two sections to Helmingham school at its foundation: the Upper School and the Lower School. The distinction was not an academic one. The Upper School was for the sons of farmers and tradesmen of professional people. The Lower School was for the children of farm workers.'[131]

Today, this sounds so wrong, but of course the privileged few still benefit from a private education while the rest of us have to accept a place at a local state school. I came close to having a private education myself because, after I failed my eleven-plus, my father entered me for the Common Entrance exam a year later. His employer, Barclays Bank, had a bursary scheme that funded a few places each year for children of its staff. I was one of three interviewed for a place at an Oxford boarding school that had two bursary-funded places. I was the one who missed out and often wonder how my life might have turned out had I been privately educated.

Years later, I realised that the private school I should have attended, which my father could probably have afforded, was on my very doorstep in Leiston, where we had moved to when I was thirteen years old. Summerhill was everything that the secondary modern school was not: liberal and exciting, with students drawn from across the world. Occupying what had been the home of members of the Garrett family, Summerhill School moved to Leiston in 1927, having been founded by A. S. Neill six years earlier in Hellerau, a suburb of Dresden.

Neill's parents were teachers and strict Calvinists, and after qualifying as a teacher himself, Neill began to question the disciplinarian approach to education that was dominant in the early years of the twentieth century. Neill felt that children were innately good, and that children naturally became just and virtuous when allowed to grow without adult imposition of morality. Summerhill reflected Neill's anti-authoritarian beliefs and the children, not the teachers, made the rules. But what gave Summerhill something of a reputation when I was growing up was that lessons were optional; children learned when they wanted to, not when they were told.

For a reason I cannot now remember, Zoe Readhead, A. S. Neill's daughter, showed me round the school in 2015 and one incident stuck in my mind. As we toured the school, we chanced upon a young lad who was talking with someone, I assume his mother, via video on his iPad, using the school's Wi-Fi. In every other school I've visited, he would have been told off for skipping his class, but Zoe simply looked over his shoulder and said 'hi' to the person on the boy's screen and our tour continued.

As you might imagine, OFSTED found it difficult to inspect a school where lessons were non-compulsory and in 1999 Summerhill was issued with a notice of complaint over the policy of non-compulsory lessons. The entire school contested the notice and four days into the court hearing, the government body's case collapsed. Had the case been lost, the school could have closed, or perhaps worse, lost the philosophy that had successful turned so many often troubled youngsters into well-rounded, socially mature citizens. Summerhill is as much a therapeutic community as it is a school, and it came as no surprise to learn that one of the most interesting and engaging people I've ever met, who for a time attended the Norwich Quaker Meeting, happened to have gone to Summerhill.

Having failed my eleven-plus, and then failed the interview to win a place to that Oxford boarding school, I went through my secondary school years believing that I was a failure, and so, while I attended lessons, I learned little and only decades later realised that my problem was that I was very bright, not very stupid. I'm sure that I would have thrived if encouraged to chart my own path through the curriculum rather than be forced to sit through uninspiring lessons taught by teachers who themselves felt second-rate because they were not teaching at a grammar school.

Having trained as a teacher himself, and being married to a headmistress, George Ewart Evans retained an interest in education throughout his life. I'm sure he would have found Summerhill intriguing, and living for several years at Blaxhall, a few miles from Leiston, he must have known of the school and perhaps, with his writer's inquisitiveness, even met A. S. Neill. Both men were willing to challenge convention, one setting up a school and, according to historian Alun Howkins, the other for a time a member of the Communist Party and keen to see a world free from oppression.

While living at Helmingham, Evans got to know Dan Pilgrim, who was born in 1882 and so had started school at Easter in 1887, at the age of five. 'He was,' wrote Evans, 'a very bright pupil; yet, though the 1880 Act compelled school authorities to enforce attendance up to the age of fourteen, Dan Pilgrim left two or three years early.'[132] There was pressure then on children to start work as young as possible, to bring more money into the household. 'He'd passed an exam and had a Labour Certificate testifying that he had reached the required standard of education.'[133]

Today, most parents want their child to get as good an education as possible and around half go on to university,[134] even though this means starting your career with a significant student debt. But Pilgrim grew up at a time of an acute farming depression that started around the time he was born and lasted until the First World War. 'This meant farmers more than ever wanted child-labour, and the parents were in no position to refuse. Most of the farm workers were day men, that is they were paid by the day; and often in the winter season they took home only three or four days' wages.'[135]

At Earl Soham, a village six miles from Helmingham, Charles Last, who was born in 1878, felt less encouraged by his time at school, explaining to Evans: 'I went to school at Earl Soham and I paid tuppence a week for going there. I left when I was eleven one Friday afternoon; and by eleven o'clock on Monday morning I was away. I was away with a farmer down by Stonham church, rooting out docks with a little two tined fork. The schoolmaster – well he'd got my tuppences and I reckon he didn't care what happened to me afterwards.'[136]

There was little help for those who found themselves without enough work to cover the family outgoings. It was not until 1911 that Lloyd George's Liberal government passed the

National Insurance Act that created an unemployment fund. It only paid seven shillings a week and took a while to get going because you could only draw dole money for one fifth of the length of time you'd been paying in. In other words you had to have been contributing for five months to qualify for just one month's payments. For all the criticism it receives, the benefits system has vastly improved over the last hundred years.

Although Dan Pilgrim left school at the age of eleven, his education did not stop there. It was reassuring to read that 'I went to night school after I left school. I attended for three or four years and I won four or five prizes there. The school was held in the Mission Hall at Helmingham. I had to help my father to read and write so that he could go there. He took a farm and he wanted to learn to read and so on.'[137] His was perhaps the first generation to grapple with literacy, something today we take for granted.

As a writer, literacy was at the core of Evans's life and language was at the heart of much of his research. As Gareth Williams, his biographer, explained: 'Ewart Evans was the grammar-school stickler for the correct use of English, whose pleasure in the adoption by the syndics of the Cambridge Examination Board of *Ask the Fellows* as an A-Level text stemmed from far more than self-interest.'[138] His love of history, which dated back to childhood, encouraged him: 'in his advocacy, via regular contributions to *The Times Educational Supplement*, of family history projects for schoolchildren and the educative role of museums in stimulating a lasting historical curiosity and awareness, he was well in advance of current thinking.'[139]

Despite this strict determination to use English correctly, Evans went to great lengths to understand the roots of the vocabulary used by the Suffolk people who told him their stories. As his biographer Gareth Williams explained, 'He believed dialect words and phrases ought to be related to the context of their use, that language was conditioned by its environment.'[140] Having grown up myself with people who spoke with broad Suffolk accents and used many of the dialect words, I doubt this will have been easy. I can remember people describing a storm as a tempest and a thrush as a mavis, both words familiar to Evans from his reading of early poets 'like Chaucer, Spenser and Shakespeare'.[141]

Thinking about my wife's family, who had been farming for centuries and probably never ventured far from home, it's easy to see how the words used to describe things had been passed down through the generations. Evans recognised this too: 'they were all brought up in contact with the land, and used language that was steeped in centuries of continuous usage, describing processes and situations and customs that had gone on uninterrupted since farming began.'[142] Although my wife's great-grandfather had left the family farm in Cambridgeshire and moved to Theberton in Suffolk, three of his sons had farmed all their lives within a few miles of where they were born.

If a Victorian farm worker had moved to another farm, he would have had to move his few pieces of furniture with horse and wagon to his new tied cottage. In 1961, when my father moved from Brantham on the Essex–Suffolk border to Needham Market, a distance of just sixteen miles, we too moved house and I had to change school. Yet for the final few career moves I made, moving house was never considered and I thought nothing of driving more than forty miles each morning from my house to the office where I was based.

Today, paradoxically, I travel far fewer miles than I once did, but have Zoom calls with people all around the world almost every day. I see this as a trend that will continue, particularly as concern about climate change grows. Extensive travel will, I think, be remembered as a phenomenon of life in the second half of the twentieth century that emerged as cars become more affordable in the 1960s and declined when people became comfortable with the convenience of meeting online. The impact this trend towards global conversation has on the evolution of the English language will undoubtedly be the subject of future research.

Historians come in all shapes and sizes. Some faithfully and accurately describe the past they have researched in dense, heavily referenced prose that daunts all but the most determined scholar. Other use storytelling to liberally interpret the way things were, and then explain them in ways anyone can understand. Evans, I think, struck a happy balance between accuracy and accessibility. His writing was factual, and faithfully recorded what people told him. But it was also very readable, although, unlike another Suffolk writer, Ronald Blythe, he never felt the need to fictionalise his work.

Ronald Blythe's best-known book, *Akenfield*, illustrates well the difference in approach adopted by the two writers. Gareth Williams summed it up: 'The village was an occupational community, not a setting for a smock and straw whimsy. For this reason, the acclaim accorded to Ronald Blythe's portrait of an East Anglian village, *Akenfield* (1969), given the soft focus in an equally famous film by Peter Hall, raised George Ewart Evans to a pitch of indignation.'[143] Both Blythe's book and the film quickly became popular and gained positive press reviews, and the book remains in print forty years later. Perhaps Evans felt that it was fiction masquerading as fact. What united the two writers was their wish to give us all an insight into life as it once was in rural Suffolk.

I'm sure those who knew the two men will agree that while Evans was an introvert, Blythe was less reserved and no doubt enjoyed appearing in a cameo role, as the village rector, in Peter Hall's film. As a lay canon at Bury St Edmunds Cathedral and a reader in his village church, he will have felt at home in cassock and surplice.

Evans, on the other hand, appeared slightly uncomfortable on screen in the BBC film *A Writer's Suffolk*, which he wrote and narrated in 1980. One scene comes to mind in which he is talking with a man who is leading a horse and loaded wagon from a cornfield. Even though standing in a field, Evans is still wearing a jacket and tie, which with his rather clipped received pronunciation meant that he did not appear at one with his surroundings. But of course, even forty years ago, formal dress and speech were more common than today, and perhaps I'm wrong to draw conclusions based on today's societal norms.

Both Blythe and Evans were masters of the art of storytelling, with Blythe perhaps more willing to create material to arouse the curiosity and hold the interest of his reader. Evans, on the other hand, strove to inform rather than entertain. I think my approach differs from both writers, revealing more about my own life, fears and feelings than either of them chose to do. That, I sense, is more important today than it was fifty years ago; the readiness with which people today share their lives on social media creates an expectation that authors will similarly expose their innermost selves.

Storytelling has always been used to educate, inform, influence and encourage us to act in this way or that. The Bible is really a collection of stories, written at a time when very few were literate and word of mouth the only way to share the message each story was trying to communicate. The better preachers tell stories, rather than bang the obedience drum, and my earliest memories of school are of sitting in a circle on the floor, while the teacher read us stories. Storytelling paints pictures in our minds, often anchored by strong visual metaphor. For example, when in 1945 Churchill described the chill that defined postwar relations with Russia as an 'iron curtain', he coined a phrase that even seventy-five years later remains familiar.

A form of storytelling for which Blaxhall is famous today is folk singing. The Ship Inn dates back to 1700 and has seen generations of local families singing traditional songs and step dancing. Joe Row, born in 1873, was one of the oldest Blaxhall residents whom Evans interviewed when writing *Ask the Fellows*. He died in 1956, the year the book was published. For an oral historian, he was something of a find, because 'his is one of the oldest families in the village'.[144] In the eighteenth century, his family had been landowners, but since the beginning of the nineteenth century most of the men in his ancestry had earned their living as farm workers.

Joe was one of the village's regular folk singers. When Evans got to know him, 'he had been singing one traditional song at the Ship for well over fifty years.'[145] Inevitably influenced by George Ewart Evans, the BBC shot a short film titled *Here's A Health to the Barley Mow* at the Ship in 1955, and the folk-singing tradition continues to feature in pub life to this day. Writing about the film, Evans explained that 'the folk in this village are accustomed to improvise the words to the old folk songs.'[146] I guess the songs would evolve over time, with new characters and incidents introduced, making them something of a commentary on life and its challenges.

There is, I was told, usually a folk-singing session in the Ship on most Monday afternoons, where villagers have always gathered after a morning spent at the weekly market at Campsea Ashe, two and half miles – so a comfortable bike ride – away from Blaxhall. Cattle and sheep are no longer sold at Campsea Ashe, but the habit of going to market each Monday morning is deeply ingrained, and browsing the bric-a-brac at the Clarke & Simpson auction somewhat addictive. I can remember taking cattle and pigs to the market here with my father-in-law Michael in the 1970s and it was also the usual destination for unwanted bull calves born to the Jersey cows at Blackheath Estate.

Interested in the way this tradition had survived, I spoke with Dr Diane Keeble-Ramsey, a senior lecturer at Anglia Ruskin University who teaches organisational behaviour. Diane told me that she was the last Blaxhall-born singer, and alongside Robert Savage's granddaughter Daphne Gant, whom I had also met, one of the stalwarts of the local folk-singing scene. She explained how the songs were never written down, but passed down from one generation to the next. Furthermore, each singer had their own song, and it was considered bad form to sing someone else's. I guess that in years gone by, the bar at the Ship became something of an educational jukebox, with a pint the price you'd pay to hear a story of local history sung.

Our ability to sing and make music, both for entertainment and to tell stories that encourage social action, sets us apart from any other species on the planet. There is, I think, something deep within us that allows us to connect through music, song and books. I can remember learning to play the recorder at primary school and for the past few years I've been having piano lessons. People still learn to play musical instruments or sing, despite that fact that Spotify means you can hear almost anything at the touch of a button on your phone.

Church

One place where music and singing carries real significance is in church. The chapels of Wales are famed for their choirs, and some of the greatest oratorios, by composers such as Handel and Bach, were commissioned by the Church. Research published by Oxford University has shown that group singing, as happens in a church service, strengthens social bonds, reduces stress and improves our sense of wellbeing.

Bringing people together, uniting them in reciting familiar prayers and singing hymns, has proved over the years to be a very effective way of keeping people where you want them. Etymologists date the phrase 'God-fearing' back to the twelfth century and I have little doubt that early Christians stuck together and sang their hearts out because they were afraid of incurring God's wrath. Christianity has long used the threat of eternal purgatory to control our behaviour through life.

Fear is a powerful motivator, but music has been effective at encouraging people to remain faithful to the church. As Evans eloquently put it: 'Religion is also important for the central part it has displayed down the ages in art. Painters, sculptors, poets, writers, musicians and actors could all be cited as witnesses to the unworldly, undefinable part of their existence and the part it has played in their inspiration.'[147] The Church was one of very few organisations able to sponsor artists, and so inevitably much creativity was biblically focused.

In rural East Suffolk, religion has had a more subtle, but equally assertive influence over the lives of all within each parish. In many villages, the largest and often oldest building is usually the church, and the Act of Uniformity, passed in 1558, made both attendance every Sunday compulsory and also adherence to the Book of Common Prayer. Going to church was simply part of the weekly routine, welcomed because Sunday was usually the only day you didn't have to work.

In sixteenth-century England, you could be fined one shilling if you did not show up on a Sunday or holy day. A shilling was for most people then a day's wages so people would think twice before skipping church. Life was equally tough for the clergy, who could be imprisoned for life, or worse, if they deviated from the Book of Common Prayer. This book was first published in 1549 and remains in use today, although many now prefer to use a newer book called Common Worship, which appeared in the year 2000. Life imprisonment for using the wrong prayer book seems harsh, until you remember that, previously, dissenting clergy had been publicly burned at the stake, with, if they were lucky, a keg of gunpowder tied round their necks to shorten their suffering in the flames. In Suffolk, most burnings took place on Cornhill in Ipswich, which remains to this day a place where any public protests tend to be staged.

Organised religion has quite sensibly incorporated customs and traditions that pre-date Christianity. The Christmas celebrations were conveniently timed to coincide with the winter solstice that was already a cause for partying, as it marked the turning point between the dark days of winter and the lengthening days that heralded spring. The ancient Norse people worshipped the sun and thought it responsible for the changing seasons. They used the word yule to describe what they saw as the cyclic nature of the sun's influence. Bringing mistletoe indoors was another pre-Christian tradition, and centuries later it was Coca Cola who made popular the image of a round, jolly Saint Nicholas, dressed in red with a flowing white beard.

I think that George Ewart Evans described himself as an atheist, although being married to a practising Quaker, and allowing his children to be educated at a Quaker school, he was clearly respectful and tolerant of faith. Indeed he was very aware of the role religion had played in informing the customs and behaviours he made it his calling to study. 'There are other reasons why religion should not be ostracised as a topic for consideration by oral history,' he wrote. 'Apart from doctrinal adherence to a particular religion, it should be recognised as a current form of what are now classified as myths.'[148]

He was also very aware of the way that many of the traditions one associates with, in particular, the Church of England, pre-date the introduction of Christianity to Britain. 'And the development or evolution of various religions shows the incorporation of many of those earlier beliefs and practices into a later context.'[149] The Suffolk he researched and wrote about had an innate connection with the soil, the seasons and the reliance people had on a good harvest if they were to endure the winter without going hungry. That link with the land was slipping when Evans was writing in the 1950s and 1960s, and has now almost completely been lost.

I discussed this change with Brian Chester, whom I had first known in the late 1980s when I was a marketing manager in the fertiliser industry and he edited *Anglia Farmer,* a local farming magazine. Now long retired, Brian had lived in Hoxne for forty years and in 2011 had been appointed a lay canon at Bury St Edmunds Cathedral in appreciation of his many years chairing the Hoxne Deanery. For a long time, Brian advised Suffolk clergy on agricultural matters, helping to reunite the church with the farming community.

He reminded me of the church festivals still celebrated that have a connection with the land, for example Rogation Sunday in late spring, when vicar and congregation leave the church to 'beat the bounds' (walk the parish boundary), blessing the growing crops along the way. Lammas is a summertime celebration of the first bread made from the newly harvested wheat, and later there is the harvest festival, which most will be familiar with. Finally Brian told me about Plough Sunday, which occurs in mid-winter when a plough would traditionally have been taken into the church and blessed.

This would be difficult today as modern ploughs would not fit through the church door, but there is usually someone in a village with an old plough that can figuratively represent winter ploughing. Rachel Cornish, the vicar of Blaxhall and surrounding parishes, told me that some of these traditional links with the soil were still celebrated in one or two of the churches within her benefice.

The old Suffolk people Evans met and interviewed for his books may have been privately sceptical about the power of religion over nature, but most erred on the side of caution and put in an appearance at church, at least when reminded of the possible relationship between God and their yields of wheat, barley and oats. As Evans pointed out: 'For a long time I was puzzled by the crowded churches in East Anglian villages to celebrate the harvest. Invariably the church was well decorated and instead of having a sprinkling of worshippers as it had for most of the year, it was crowded.'[150] Furthermore, while women attended church more regularly than their menfolk, at the harvest festival service, there 'was a far bigger proportion of men among the congregation. It was clear that the old pre-Christian impulse to hold a ceremony at harvest had still been sustained.'[151]

Victorian landowners expected their staff, both on the farm and in the house, to attend church. Although not formally allocated, the best pews in church were always occupied by the better-off families in a village, and the pews where workers and their families were expected to sit were usually in easy view of where their employers sat. Questions would be asked on Monday if you had not been seen in church the morning before.

In many communities until the middle of the last century, the church very much represented the establishment. As we have already noted, some of the finest churches in East Anglia were built by men who had become very wealthy from the wool trade with the Low Countries. In many estate villages you inevitably find the church situated next to the hall, rather than in the centre of the village where most people lived. Even in the late 1960s when I was an altar server and so a regular attender myself, everyone seemed always to know who the landowners were and where in the church they habitually sat. They will have first come to church with their parents, who will have done the same, so often a family will have occupied the same pew for centuries.

Always sitting in the same place is habit and not unusual, although Evans made a point of noting the hierarchical positioning of families in churches. When I go to my Quaker Meeting I too always sit in the same place, as do most of the other regular Friends and attenders. My preferred seat gives me a view through one of the high windows of the trees outside. It is also in a corner, against the wall, so I do not have people sitting behind me, which I find oddly disconcerting.

I'm not sure how my Quakerism would have been viewed

had I been an estate worker a century ago. Differences in Christian interpretation have down the years resulted in the evolution of an array of broadly similar faith groups, each quite unique in its style of worship and degree of adherence to the Bible. By definition, Methodists are methodical in their beliefs, Roman Catholics perhaps the most ritualistic and Quakers the most liberal, preferring to refer to the Light than make any direct reference to God. I like this pragmatic approach, as it seems a little too convenient to bundle together all we don't understand about life and death and attribute it to a mystical bearded figure.

Most towns and many villages have a number of chapels and Meeting Houses as well as the Church of England parish church, although this is usually the oldest and largest religious building. Most were originally Catholic, only changing when Henry VIII found that religion's rules got in the way of his wish to remarry, prompting him to break with Rome. Happy Sturgeon, born in 1917, had both an unusual name and did not worship in the same church as her employer, telling Evans: 'The other girl had the morning off as she sometimes went to the Church of England; but me being Congregational, that didn't count as Christianity; I didn't get time off for going.'[152]

At some point in 1961, George Ewart Evans's experience of religion and mine all but overlapped. I remember it clearly because of what remains one of the most embarrassing moments of my life. The incident was not in itself significant, but as a shy six-year-old, getting noticed by anyone was something I tried at all times to avoid. It took place in Needham Market parish church, which as I later discovered, was just across the road from where Evans and his wife, my then school headmistress, lived.

We had been living at Needham Market for a few months when my father developed an interest in religion. His parents, my grandparents, had been enthusiastic Methodists and I knew that my father had gone to Sunday school when a child, but now he felt drawn to the Church of England. On most Sunday mornings he would insist I wore my school blazer, shorts and cap, and, with my mother and brother, accompany him to the church, which was on the high street, some 200 yards from our bungalow. The vicar, Rev. W. G. Hargrave Thomas, was a black-cloaked, larger-than-life figure and I suspect it was his eccentricity that drew my father, just as I now also find unusual people compelling company.

My moment of acute embarrassment came one Sunday morning when a large elderly lady sitting in the pew behind us leaned forward and, looking at me, said to my mother, 'Hasn't he got a lovely singing voice; he should be in the choir.' From that moment to this, I have never, ever sung anywhere, in any situation, social or otherwise. When confronted by a hymn book at a wedding or funeral, I will either mime if I think others will not notice, or simply keep my mouth tightly closed. This is another reason why the Quaker style of silent worship suits me.

I have vague memories of that same vicar visiting the school, because as Evans wrote in *From Mouths of Men,* Hargrave Thomas had been a school manager, and so often visited the school. He was I think one of those rather eccentric clergymen not unusual in the Church of England before the era of political correctness and safeguarding. A friend of mine who is an ordained Anglican priest told me that there is little room for individuality in the twenty-first-century Church of England. Perhaps this is one of the reasons why, for most people, going to church has lost its appeal.

Time magazine described Hargrave Thomas as one 'in the best British tradition of unconventional vicars'.[153] The article went on to explain how, because he found most hymn tunes boring and hard to sing, he brightened them up by changing the music to something more joyful. For example, he had dressed up the old Anglican hymn 'Rest of the Weary' with the syncopated melody of a British bandleader's current theme song, 'Here's to the Next Time'. It's easy to see why my father started going to church, something he continued to do until he died. At his funeral his then vicar rather tellingly said that my father had been a better friend to him than he himself had been to my father.

Hargrave Thomas would not survive as a parish priest in the current century, but sixty years ago his unusual behaviour was accepted, even cherished, by his faithful flock. In his autobiography Evans described how his first encounters with the vicar had left an impression: 'He was wearing his black, wide-brimmed "Devil-Dodger" hat. When I looked closer I could see two pairs of little shoes each side of a pair of hob-nailed boots. When he got nearer he opened up the cloak and two little boys he had enfolded on their way to school darted out merrily.'[154]

My parents taking me to church, the birth of my sister in 1964 and those years at primary school mean I have many memories of my time living at Needham Market. Not all of these were good, and some gave glimpses of what might lie ahead as I progressed through my life. I can also remember how my father's parents would visit us quite often, driving the twenty-five miles from their home in Colchester in my grandfather Ernest's 1946 Hillman Minx. Born in 1893, he was nearing retirement before he bought his first car, I think so that he could visit us more easily.

He had set off from our bungalow for home after tea one summer Saturday evening, I think in 1963, and turning right onto Needham Market High Street had not seen a car heading westwards and so there was a collision for which he was clearly to blame. The local garage collected his car, and he and my grandmother finally went home on the train. I'm not sure if the car was ever repaired, but he never drove again and the next year he developed bowel cancer and died. At the time, he seemed to me to be an old man, but was just seventy-one when he died, which now that I myself am just a few years short of that age, does not seem old at all.

The scene of my grandfather's accident was, I now realise, just a few yards from where Evans had then been living. He too had noticed an increase in the volume of traffic passing through Needham Market, writing that: 'We lived on the main street, a busy road that took all the traffic from the coast to the Midlands.'[155] It was a few years later, in the 1970s, that what is now the A14 was built, by-passing Needham Market, Stowmarket and Claydon. Later, in the early 1980s, Ipswich was also bypassed, with the Orwell Bridge taking traffic over the river and straight to Felixstowe. It would be unthinkable today for the main road to pass through the towns and villages as it once did.

Just as I was introduced to religion as a boy, so too was Evans, although, being Welsh, he attended chapel rather than church. His father William had been a deacon at the Calfaria Baptist chapel next door to the shop where the family lived. He had also been superintendent of the Sunday school, so Evans will have had a far more religious upbringing than me. Although, like many who are raised in a religious family, he moved away from religion in adulthood, he did not stray as far from his chapel roots as one might think.

When a supply teacher in London, he 'sampled two of the usual diversions of a Thirties valley boy going up to London for what was effectively the first time. These were a visit to a Welsh chapel and a Sunday evening experience at Marble Arch, where the Welsh used to gather in a good-sized crowd.'[156] 'Many came up to the Welsh corner just to sing and as it was Sunday night most of the singing was of Welsh hymns.'[157] I guess it was the singing alongside fellow Welshmen that drew him there, just as we all tend to seek out our fellow countrymen when abroad and alone. He also attended the Welsh chapel at Castle Street, near Upper Regent Street, a few times, but because he 'wore flannels out of necessity while the rest of the congregation were in conventional Bible black',[158] he attracted some stern looks from the regulars.

Whatever your views on religion, you cannot deny the long connection that exists between religion and charity. As well as promising the faithful the dubious delight of a life of sorts after death, much charitable activity has for centuries been inspired by faith. It was as long ago as 1130 that Frenchman Henry of Blois, then Bishop of Winchester, founded the Hospital of St Cross, a place that offered bed and board to thirteen people who lacked the means to support themselves. Today there are twenty-five residents, all retired, single or widowed men, who are called brothers, and who attend church each morning and show visitors round the ancient buildings.

A similar institution, the Great Hospital, sits next to Norwich Cathedral and since 1249 has been accommodating 'aged priests, poor scholars, and sick and hungry paupers,'[159] which today is translated as being of retirement age and entitled to housing benefit. Unlike Winchester's Hospital of St Cross, Norwich's Great Hospital accepts women as well as men and there is no expectation that residents will attend church services. I've visited the Great Hospital many times over the past thirty years, as it often hosts charity dinners and even business breakfasts, particularly those organised to raise awareness of, and funds for, other good works in the city.

Norwich's cathedral close, a forty-four-acre riverside estate of eighty highly desirable, attractive and often ancient houses and flats, is more commercial, with the Dean and Chapter charging market rents to those wishing to live in what might be described as a heavenly and very convenient city-centre location. But then maintaining the 900-year-old cathedral costs money and a team of seventy are employed to run the place. It may not be a charity, but its income is spent on furthering the work of the cathedral.

In late-Victorian Suffolk it was the landowners and clergy who were the respected and revered figures of authority in a village. They had held power over people for centuries, and were usually the men (they were always men) who endowed and ran local charities, and so they were the people you went to cap in hand when times were hard and your children were hungry. You see aspects of this even today, with many parish councils still chaired by local landowners, to whom the other members defer and whom few dare to contradict.

But as the twentieth century progressed, these traditional figures of authority found themselves demoted in public esteem and replaced as celebrity culture took over. Captain Sir Tom Moore illustrated this point well, when, having set out in 2020 at the age of one hundred to raise £1,000 for the NHS by being sponsored to walk up and down his garden, his endeavour captured the public imagination and he raised a staggering £38.9 million. His legacy is the Tom Moore Foundation, a charity that continues to raise money in his name.

The celebrities of the past were those whose notoriety meant that everyone was talking about them. In the centuries before radio and television, when not everybody could read, it was usually those who had done terrible things who became famous, particularly if it was something people could relate to, like robbery or murder. Dick Turpin, who became a feared highwayman until he was caught and hanged at York Castle in 1739 when not yet forty years old, became a legend, as I suspect did Suffolk boy Toby Gill, who was suspended in a gibbet at Hinton crossroads, between Blythburgh and Westleton, just eleven miles from Blaxhall, in 1750 as punishment for a particularly unsavoury murder. While hanging someone from a rope quickly killed them, in the steel cage of a gibbet they might last days before giving up the ghost. The car park at the spot where Toby died is to this day known as Toby's Walk, although few who stop there know the reason why.

Perhaps the difference between those respected in the past, or whose notoriety brought them into the public eye, will all have had a connection, or at least been familiar with the established church. Today, it is more likely that those whom people choose to admire and follow have no connection with the church at all.

There were sound practical reasons why the local landowner and clergymen had such a high level of control over their neighbours. From the Middle Ages, churches levied taxes known as tithes on their parishioners, taking 10 per cent of the harvest each year, storing it in a tithe barn and selling it to fund the vicar's pay and the upkeep of the church. This practice can be traced back to the Bible, which encouraged early Christians to willingly give to the church because 'God loves a cheerful giver'. I'm not sure how many, when times were tough, gave cheerfully. Even today, some of the more evangelical churches are funded in this way by their congregations.

Later, from 1601, a compulsory poor rate was levied by each parish, with the sum each homeowner paid based on the value of their house. Money collected from this local tax was held by the church and distributed to the poor and needy as the village overseers saw fit. These overseers were unpaid and appointed by the vestry, which today would be called the parochial church council. The vestry was, inevitably, chaired by the vicar, with other members being the churchwardens and local landowners. The overseers also had responsibility for seeing that orphans were cared for, as Evans wrote: 'The children of parents who had died or were too poor to keep them were looked after by overseers.'[160]

The system of local taxation is not that different today, with each district, city or borough council collecting rates that are still today linked to the value of our house, and running homes for orphaned children and the elderly. However, those who control the level of taxation and how the income is spent are now democratically elected rather than appointed by the church. Many might consider this more secular approach preferable, but today success in an election is as often as not dependent on which political party a candidate belongs to. Individual merit seems to matter little, because providing you can secure nomination by whichever political party holds local control, election to office is in vast swathes of East Anglia a mere formality.

Doing good in a community is not, and never has been, the sole preserve of the church or those with land, although obviously the greater your wealth, the more time you can devote to helping others. Villages such as Blaxhall were largely self-contained. Most people did not have savings, but lived from week to week spending the money they earned in the village on food and services that were provided by others in the village. Bartering was commonplace and everyone did what they could to help each other out. The phrase 'charity begins at home' comes to mind, because in Blaxhall as I am sure in many rural villages, there were few families and much intermarriage, and so the connection between people was deep and strong.

I suspect that the Second World War marked the turning point, with everybody pulling together while it was being fought. The years that followed saw cars, televisions, supermarkets and far greater awareness of the world beyond the community within which we live. It was no surprise to learn that Evans's family played a part in settling children and young mothers who sought safety in his Cambridgeshire village from the German bombers that were destroying much of London. 'Trains of evacuees now followed one another out of London. A train of mothers and children, young ones and others of school age, deposited a large party in the village of Shelford. I was among those who helped receive them at the village hall. This was for me one of the most affecting experiences of the war,'[161] Evans wrote.

One lucky evacuee from London stayed with the Evans family. 'At home we had an evacuee schoolboy from London, a pleasant ten-year-old who fitted into our household quite smoothly.'[162] It was perhaps this daily reminder of the war, even when far from the frontline or bombing, that meant that everybody felt involved. The global pandemic that struck in 2020 rekindled some of that community spirit, with volunteers shopping for the vulnerable and old, or driving people for their vaccination. The difference being that, discouraged as we were from mixing with neighbours, most social exchange was online.

Trade

We have a set of brass scales in our living room. They're quite large and once sat on the counter of Blythburgh village shop, which my wife's grandfather Bernard Hawkes took over from his wife's aunt Alice in the 1950s. The scales have a set of brass weights ranging from seven pounds down to a quarter of an ounce. When he retired, the shop closed and he gave the scales to his granddaughter.

Bernard grew up in a family that had several shops, each selling sweets from the factory his father and uncle ran in New Street, Chelmsford. Hawkes Brothers supplied many other shops too, delivering sweets around Essex in an open-fronted sign-written van. They also had a contract with BOAC, supplying the humbugs that air stewardesses handed out before take-off and landing. Sucking on a boiled sweet as the plane took off was the recommended way to stop your ears popping as the air pressure changed.

I'd bet that William Evans, father of George Ewart Evans, also had a set of brass scales in his shop next door to the Calfaria Baptist chapel in Glancynon Street, Abercynon. Many lines were sold loose in the early years of the twentieth century and carefully weighed out into paper bags by the shopkeeper or his assistant. In the late 1940s the introduction of self-service stores, where the shopper filled a basket themselves and took it to the counter to pay, meant that food had to be pre-packed, and over time counter-top scales became redundant.

Convenience and bringing the consumer lower prices saw the rise of the supermarket and the subsequent closure of so many village shops. When I was growing up in Leiston in the late 1960s, my mother would walk up the hill to the International Stores to buy most of the groceries she needed, despite there being a smaller, but more expensive grocer's shop almost next door to the bank above which we lived. She did not mind carrying her shopping a hundred yards home as she had learned from her mother the importance of watching the pennies. Both shops are now long gone, replaced by a large Co-op supermarket a little further along Sizewell Road. It was built on the site of the smaller Co-op shop it replaced, and its adjacent distribution depot. The Co-op now has a large car park as most people now drive, rather than walk, to do their shopping.

But while supermarkets have got bigger, with many now stocking more than 30,000 different lines, the village shop is enjoying something of a renaissance. The Plunkett Foundation is a charity that helps rural communities work together to open and run businesses such as shops and pubs. Village shops have been closing over the past few decades, often when the shop's owner retires. The Plunkett Foundation supports a network of more than 370 community-owned shops, some very small, perhaps based in a converted portable building, and others existing shops that have been taken over by local people using a community share issue to raise the bulk of the capital cost.

Plunkett recognise that a community-owned shop, staffed by volunteers with perhaps a part-time paid manager, does far more for a rural village than simply provide a handy place where you can buy milk, eggs and perhaps a local paper. I helped one village, Little Plumstead in Norfolk, through the process of a community share issue, seeking grants and even identifying a company able to build a shop within the budget they had available. Situated in what was a derelict walled garden, amidst an estate of new houses, it has a café as well as a shop and has become a popular place for neighbours to meet.

A hundred years ago the Suffolk village of Metfield supported six shops, including a draper, butcher and fishmonger, as well as a brick kiln and three corn mills. By 2005 the one remaining shop in the village had closed and, unhappy with this, a small group of local residents got together to try to find a way to turn it into a community shop. The building was put up for sale and bought at auction by a generous member of the group who then leased the ground-floor shop to a newly formed Metfield Stores Community Interest Company. Some volunteers from the 400 or so residents of Metfield and from nearby villages formed a board of directors. Many more helped to refurbish what had become a tired and very run-down shop, and then went on to become sales staff, led by a part-time manager.

When I visited, to chat with Bridget Morley and Jan Rusted, both directors of the CIC, the shop looked as smart as it must have done in its heyday, painted in a bright pastel blue with large windows that draw the eye to the array of treats and essential supplies inside. The shop's governance document states their intention to use 'suppliers on a local level to support food production in the area', and wherever possible this has been done. Bridget and Jan told me how during the recent pandemic the shop had been something of a godsend, especially for the older residents who preferred not to travel five miles to nearby Harleston to do their shopping.

Today people will think nothing of driving into Harleston, with its far larger population and a good range of shops to stock up their larders. But in the days before people had cars, most would have shopped in the village. Walking, or getting a lift on a passing horse and wagon, into Harleston would have been a special occasion, and to venture sixteen miles further to Norwich would have entailed catching a train at Harleston, then changing at Tivetshall for the journey north to the city. Trains still pass through Tivetshall as it's on the main Norwich-to-London line, but the station, and indeed the whole Waveney Valley line to Beccles, closed in 1966.

Another community-owned venture that caught my eye was a shop and café in Badingham, a particularly rural village four miles north of Framlingham. I arranged to meet Malcolm Knott, one of the directors of this community interest company, and set off one day for the hour's drive from my home to the shop. Within minutes of entering Suffolk, I encountered a closed road at Syleham and, following a white van, took a narrow lane to circumnavigate the blockage. At the first junction, where my satnav told me we had to turn right, a lorry was blocking the road, unloading pallets of bricks with a crane. This delay, and the rest of my journey, along increasingly twisting narrow lanes, made me very aware of just how rural Badingham is.

The shop and café occupy the former stables of a house on Low Street, one of the village's two roads. A post box and bus stop are nearby, although I doubt many buses pass through the village. Malcolm explained how a good neighbour had allowed the group to convert the stables into the shop and café. With outside tables and a tempting array of cakes, the café was the main reason people came. The shop sells local crafts, rather than try to stock groceries; few people live in the village and even fewer pass by so stock turnover would be low.

The postman called while I was chatting with Malcolm, and recognising one of the ladies enjoying coffee and cake, walked back to his van and returned with her post. 'I'd not got to your house yet,' he said, before jumping back into his van and continuing on his round. That innocent friendly act said much to me about the sense of community here. Not only do folk here know their neighbours, but the postman knows everyone too. I see signs of this returning in my own village, but in a very contemporary way, with Nextdoor.co.uk, a place where people find near neighbours able to help them with things.

The celebratory booklet produced in 2017 to mark the unveiling of a blue plaque outside the shop that was the childhood home of George Ewart Evans lists the businesses that were trading in Abercynon in 1910.[163] At the time the town supported eight grocer's shops, including that run by the Evans family. It also had three drapers, four bakers, two tailors and two pawnbrokers. Although many of the shops will have been small and have managed to get by on modest takings, people will have done almost all of their shopping in the town where they lived.

There is, however, a very practical reason why rural shops prospered in the early years of the twentieth century. As Mervyn Cater, who lived in Framsden, explained: 'there was only one form of public transport, the carrier's horse and cart that went once or twice a week to the town, and even then with little room or encouragement for passengers'.[164] In other words, to travel any distance was to make something of an expedition. Even today, there are just three buses a day that run between Framsden and Ipswich, and only two that can take you back home again.

Most people who live in rural Suffolk today rely on their car. Before cars and buses, going to Ipswich from Framsden meant a twenty-four-mile round trip on foot. Sam Friend, a contemporary of Mervyn Cater, who was born in 1886, talked about taking the 'hobnail express', and walking into town. Often this meant sleeping under a hedge on the way back, or, as Friend put it, 'on Mother Greenfield's pillow'. I suspect many young men spent the night sleeping under a hedge after an evening out in town. It was ever thus, so although it seems tough and even foolhardy today, those old men whose stories Evans had recorded took travelling long distances on foot quite literally in their stride.

That's not to say that people did not used to travel extensively. Suffolk farm workers regularly travelled to Burton-upon-Trent to work in the maltings over the winter, and many followed the harvest south each summer, working their way down the east coast. Read the books written by Norfolk author George Borrow, particularly his autobiographical novel *Lavengro*, first published in 1851, and you will marvel at how widely he roamed in his early years. Born at Dereham in Norfolk to a Cornish father and mother whose family had been French Huguenot exiles, he travelled extensively as a child, attending schools in Edinburgh, Norwich and Ireland. He wrote about walking from place to place and sleeping out under the stars, particularly in *Romany Rye*, in which he wrote about his time living the life of a Gypsy. His experience contrasts markedly with those Evans interviewed in Blaxhall a hundred years later, who, it seemed, had not moved many miles in generations.

There was a fascinating connection between the various tradespeople who made up a village economy, and a sound logic to how some businesses were inevitably based next to each other. For example, 'the blacksmith and wheelwright in most East Anglian villages had their shops next door to each other.'[165] This was because the blacksmith made the iron tyres that had to be heated before being placed around the newly made wheel, onto which the tyre fitted tightly when cool. It saved the wheelwright taking each wheel down the road to the blacksmiths. As Evans wrote, 'both trades were closely linked with the farm, chiefly because the horses and wagons could not be kept in position without the help of the blacksmith and the wheelwright'.[166]

In a Victorian Suffolk village, the blacksmith played an important role, creating many of the tools and implements that were essential to everyday life. This was particularly true until the second half of the nineteenth century, when increasing mechanisation meant that more and more goods were mass-produced in distant factories, rather than at the local forge. The blacksmith's shop was also a popular place to hang out, at least until the pub opened. As Clifford Race, who was born in 1898, pointed out, visits to the blacksmith were regular events. Horses that worked the land needed new shoes around every three months, while those pulling wagons on the roads needed new shoes more often, perhaps as frequently as every three weeks. Horsemen would usually wait and chat while their horse was reshod.

Perhaps the most famous blacksmith's forge is the one at Gretna Green, which as the first stop for stage coaches leaving England, became where eloping couples could marry. In the eighteenth century, you could not marry in England without parental consent until you were twenty-one. In Scotland the rules were more relaxed.

Running away to get married sounds fun, but was never something my wife and I considered. In fact, looking back, I think our marriage was arranged, as my mother-in-law went to some lengths to encourage me when I was working on the farm. I'd be fed stew and dumplings indoors on a Saturday, when the other lads on the farm had to make do with sandwiches their mothers had made for them. She was also quick to nurture any interest I showed in her elder daughter, thinking, I suspect, that the local bank manager's son was a good bet as a potential husband. In fact, it was only when she stopped trying to push us together that a romance began to flourish. Even today, more than forty years later, I cannot eat stew and dumplings without smiling at the memory of our early courtship and those Saturday lunches at the farm.

My wife's parents, in common with most farming couples, had some horse brasses hanging on their living room wall. Perhaps they were left from when the family farmed with horses, or perhaps they'd just been bought one day on a whim. Horse brasses were just something you'd see in almost any farmhouse, and they would have been made by early blacksmiths.

The National Horse Brass Society provided a clue to the popularity of horse brasses, explaining how they had originally been used not for decoration, but as talismans to protect the horses from malevolent spirits. A farmer's livelihood was dependent on his horses, so he would have wanted to protect them from the unwanted attention of anything that could harm them. Superstitious or not, few will have wanted to risk their horses' wellbeing, so most kept horse brasses. Like so many customs, the habit was maintained long after the original reason had been long forgotten.

Writing about them himself, Evans pointed out that 'the common explanation for the use of brass decorations on a horse is that they are survivals from the time when the horse was considered susceptible to the evil influences of witches.'[167] His research suggested to him that this folklore dated back to pre-Roman times, but he was pragmatic enough to note that 'Conversations with many horsemen in Suffolk has not brought any one statement about the symbolic meaning of horse brasses.'[168]

Keen to understand the connection between folklore, superstition and horse brasses, Evans explored their symbolism. Some had patterns and were quite abstract, while others more logically featured horse images. Working horses in Suffolk would wear both figurative and abstract brasses, usually mounted on a strap on their forehead or running down their muzzle. More would be mounted to a thick leather strap called a martingale that ran from the bottom of the collar to a band that ran round the horse, just behind its front legs, known as a girth strap.

Perhaps it was the mass migration from the countryside to find work in the newly steam-powered mills and factories that were springing up in towns and cities across the nation in the nineteenth century that prompted more widespread interest in horse brasses. Their popularity grew after the 1851 Great Exhibition, probably because they were by then mass-produced rather than made by local blacksmiths, so cheaper to buy, and also because they were on sale as souvenirs at Crystal Palace, where the exhibition took place. If you'd grown up in the countryside but now found yourself living in a back-to-back terraced house somewhere like Leeds, it would be only natural to want some affordable reminder of the life you'd left behind.

I can remember that when I visited Daphne and John Gant in Blaxhall, John showed me his collection of horse brasses that were hanging on the couple's sitting-room wall. John had worked all his life on the land, and Daphne's grandfather had been Robert Savage, the shepherd who featured so prominently in *Ask the Fellows*, so each of these brasses had a story that I enjoyed hearing. Most who bought horse brasses in what the National Society described as a trade that reached 'boom proportions' had simply bought them to decorate their wall, in the same way that many of my parents' generation had bought sets of three china flying mallard ducks to hang above the fireplace. Perhaps they represented the freedom young parents yearned for but that work, a mortgage and child-rearing placed out of reach.

Almost as universal as those flying ducks was the popularity of keeping a goldfish in a bowl on the sideboard. I can remember coming home from the fair as a child in the early 1960s with a goldfish in a water-filled plastic bag. Bert Stocks Fair used to come each year to the recreation ground a few yards from our bungalow in Needham Market. There was some rivalry, if both my brother and I won one at the fair, to see whose would live the longest. At least this competitive streak meant we looked after them a little better than we otherwise might have.

It was a reference Evans made to Bert Stocks Fair that made me think of those goldfish. Joe Thompson, born at Clopton in 1902, had told Evans about how he had worked at the fair through the summer of 1921. He told Evans that he had been an engine driver, operating a Burrell traction engine that was connected by a long leather belt to a dynamo that powered the fairground lights. He was then twenty years old, and perhaps enjoyed travelling round Suffolk with the fair before settling down to live and work on the Helmingham Estate. His time with the fair was in many ways like the gap year students take today before starting university.

Unlike most students on gap years though, Thompson had more, not less money in his pocket each week when he worked at the fair: 'I got five shillings a week more than a farm worker and one free meal – a cooked dinner at the middle of the day. When the fair was set up I was often an odd-job man as the engine was stationary. I used to take money at the coconut shy.'[169] Evans described Joe Thompson as being 'short in height but built as strong as an oak door'. I suspect that just as I had been swayed by my mother-in-law's lunchtime stew and dumplings, so too would a daily cooked dinner have appealed to the then single man working with the fair.

I was intrigued to find that Stocks Fair still tours East Anglia, now run by William Stocks, a direct descendant of the founder. William told me that the fair dated back to 1821 and even today, where the local council allowed it, gave goldfish as prizes. Today, though, they are kept to a very high standard, and I suspect no longer dangle tantalisingly in plastic bags along the front of the coconut shy to tempt you to try to win one. Steam engines no longer run the generators that drive the rides. Today the fair has large diesel-powered generators that are towed, along with the rides, behind lorries. You sometimes see a fair on the road, with an array of rides and equipment making slow progress, towed by a convoy of trucks.

The other lasting memory I have of the fair is the annual piano-smashing contest, although I think this was organised locally and not part of the fair's entertainment. It was simply staged at the same time as the fair was in town. Pianos fell from favour dramatically when television sets become popular in the late 1950s. To compete to smash a piano to pieces with a sledgehammer sounds strange today, but it was a serious sport when I was a child and there was a ready supply of redundant pianos waiting to be broken up. I know from my own experience that it takes years to learn to play the piano, so it is perhaps not surprising that they have fallen from favour; it's much easier to buy a TV or listen to Spotify and have entertainment at the touch of a button.

The fair was just one of many businesses that had survived for centuries. The Needham Market branch of Barclays Bank where my father worked had first opened in 1744, established by Quaker banker Samuel Alexander. Later it merged with Barclays, then closed in 2015 and today is a dentist's surgery. Other shops there and elsewhere in Suffolk will have passed down from father to son, or mother to daughter, for generations. I guess that's why so many family names are descriptive of a trade. One of my grandmothers had the maiden name Cooper; her father was a woodman near Cromer, and his father a blacksmith on the same estate. It's easy to see how a few generations back those skills could have been combined to make barrels, which of course is what a cooper did.

Another trade that has vanished from East Anglia village life is tailoring. Today we all buy ready-to-wear clothes from town retailers, who in turn often source the garments they sell from the Far East and other parts of the world where labour is plentiful and cheap. Low prices and clever advertising encourage us to buy new clothes every year, and throw away our old stuff. Few people trouble to work out how little the seamstress who makes your £12 pairs of jeans is actually paid. When you deduct VAT at 20 per cent, the retailer's margin and that of the wholesaler, then allow for the shipping halfway across the world, you realise how little money is left to pay for the labour. Many are paid less than the minimum wage for their country, which in the Philippines for example can be as little as $5 a day.

In the years before the First World War, most villages had a tailor who would make made-to-measure clothes that although relatively expensive, were treasured, patched, repaired and made to last a lifetime. In 1964 Sam Friend told Evans that he was still wearing a coat he had bought in 1911, telling him it was as good then as it was on the day he bought it, although I'm not sure everybody who knew him would have agreed. He would have been just twenty-five when he bought the coat and seventy-eight when he told Evans about it. I wonder how many twenty-five-year-olds today will still be wearing the same clothes even five years from now?

The tailor in Debenham had a reputation for making smart suits for horsemen. There was considerable status to being head horseman on a farm, and as Evans wrote: 'The farm horseman had almost as much regard for his own appearance as he had for that of his horses.'[170] It helped that horsemen were paid a little more than other farm workers, albeit not much. 'Since he had a couple of shillings more a week than the ordinary day-labourer he was able to save a little and get himself a very good suit of clothes.'[171] I suspect that because they would only wear their suit on Sundays, or at weddings or funerals, when the time came for their own funeral, the suit they would be wearing in their coffin will have been in their wardrobe for many years.

Not all village tradesmen had a shop or regular place of business, and, as there are today, there have always been travelling salesmen selling their wares door to door. The milkman is perhaps the most obvious example. Everybody bought their milk from the milkman and my generation all grew up drinking milk. Sometimes it had gone off because it had sat on the doorstep on a hot summer's morning, waiting to be brought indoors, where, in my case, for many years there was no fridge to keep it cool. We also had milk each day at school, with crates of one-third-of-a-pint bottles stacked in the playground, where it froze in winter and went off in the summer.

Fish is an equally perishable commodity that was sold door to door, and with Blaxhall being just eight miles from the beach it was inevitable that some of those Evans interviewed sold fish. As he wrote: 'The selling of fish by small traders was an important activity along the East Anglian littoral.'[172] Something I was surprised to learn was that the hamlet called Friday Street, which is near Snape, was so called because of 'the increase in traffic on that day of the week when, under the old religion, fish provided the main meal'.[173] Perhaps the name evolved as a subtle snub to those who insisted on eating fish on a Friday and the name stuck. There is a farm shop and café at Friday Street today. Next time I visit I will see if fish features on their menu.

Thomas Spindler, born at Yoxford in 1890, told Evans how he had been selling fish for nearly all his life. He was eighty-four when Evans met him in 1974 and still working in that trade. His family had worked for the Blois family at Cockfield Hall, wealthy local landowners who had owned much of the land in the area since the seventeenth century. My father-in-law had been a tenant of Sir Charles Blois as a young man, keeping a herd of dairy cows on the marshes behind Blythburgh church. The Spindlers had worked on the estate for generations, as many families did, and had at some point in their history developed an interest in the fish trade.

Just as an eldest son born into the Blois family will have grown up knowing that his destiny was to one day manage the family estate, so too was Thomas Spindler prepared from an early age to continue his family's tradition. 'My father had a little basket made for me when I was seven,' he told Evans. 'It held about 24 bloaters, and I sold them in the morning before I went to school.'[174] He went on to say that at the age of fourteen he would go to Lowestoft with his father to buy fish. For many, the interests we develop in our teens remain with us for a lifetime. I was fourteen when I started working on farms, and today own a tractor that is almost sixty years old, and so similar to one I drove in my youth. My attachment to the machine runs deeper than nostalgia, and I wonder if when I die I could be buried sitting on it, in the same way as Viking warriors were interred with their longships.

Of course this is highly impractical and rather silly, as you might think is buying lots of fish to sell in the years before refrigeration. But the enterprising Spindler family cured most of the herrings they bought, making bloaters (which are smoked whole) and kippers (which are gutted before smoking). This was as much to make them keep longer as to enhance the flavour. The advice today is to sell them within five or six days of smoking, but I suspect that in Thomas Spindler's time, they'd have lasted far longer, particularly as they would smoke many more at a time than they could possibly sell in a week. As he told Evans: 'We had two big smoke houses in the garden. We could put 10,000 fish in one.'[175]

The herrings would be carried in barrels from Lowestoft to Darsham by train, from where he would 'cart them by horse and cart from the station',[176] a journey of about two miles back to the smoke house in the garden behind the Spindlers' cottage. This was far from unusual, for as Evans observed: 'Many of the fish trading families along this coast had their own fish-curing houses in addition to their retail business. The Spindlers were no exception.'[177] While few would now buy fish cured in someone's back garden, many patronise the Aldeburgh and Orford fishermen who smoke their catch in wooden huts next to the beach. People are today far more interested in the journey the food they eat has made, with rising concern about the environmental impact of fish flown in from the Far East rather than being sourced closer to home.

Perhaps because he had spent so much time listening to people talking about how life was when horses, not tractors, tilled the soil and most food was produced close to where it was consumed, Evans was surprisingly tuned in to the climate change debate that has become one of the burning issues of the present day. It must have been as long ago as 1978 that he predicted a looming crisis at 'the prospect of the world supply of fossil-fuels drying up within the next couple of generations if consumption goes on at the present rate.'[178]

Had those Evans interviewed been alive today, they would no doubt be surprised by the growing interest in rewilding farmland. Lord Somerleyton, more familiarly known as Hugh Crossley, is rewilding a fifth of his 5,000-acre estate in north Suffolk, a process he describes as restoring the land 'to its natural uncultivated state. With naturalistic grazing systems set up to enrich biodiversity'.[179] The estate has been in his family since 1861, so it's perhaps easier for him than many, but he is by no means alone in seeking to farm in greater harmony with his natural surroundings. Agriculture accounts for 10 per cent of the country's greenhouse gas emissions, and farming less intensively, as the Mayhews have done near Bungay, or rewilding as at Somerleyton, is one way of reducing those emissions.

But while the environmental benefits of rewilding are obvious, we must not forget that farming's prime function is to provide the food we eat. In a *Sunday Times* feature, Michael Sly, who farms near Peterborough and is a director of the mustard-growing venture Condimentum, said: 'It feels like we are going back to the early Victorian period; it is a wealthy landed class re-establishing itself in the country.'[180] He has a point, because farmland can pass on to the next generation without being hit by inheritance tax.

The decades since George Ewart Evans wrote *Ask the Fellows Who Cut the Hay* have seen rural life continue to change, but recent changes suggest to me that much of what was familiar to people in the early years of the twentieth century will become equally familiar to those who live through the middle years of this one. Land ownership is changing, land management is changing, and the relentless migration of people from the countryside to our cities and towns has slowed, and may soon even begin to reverse, as new jobs are created and the benefits of living in smaller, more self-sufficient communities become more widely valued. We can become the fellows who cut the hay if we choose to return to a lifestyle that is closer to nature and to our roots.

Afterword

THERE IS A WONDERFUL Quaker tradition that at the end of a meeting, which will have been conducted largely in silence, those present are invited to share any thoughts that had come to them during worship but they had felt did not constitute ministry. Ministry is the word used to describe those things that come to you and you feel moved to stand up and say out loud during the meeting. These are known as after-words.

Florence Evans, George Ewart Evans's wife, will have been familiar with afterwords and I wonder if this was what prompted Evans to develop the habit of ending his books with a final page of reflections, that he usually titled Conclusion. It somehow feels appropriate, as I reflect on the book I have just written, to share some after-words of my own.

Although I'd been intending to write this book for decades, it was only when I enrolled on the creative writing MA at the University of East Anglia that it started to take shape. This is far from unusual, and many well-respected and highly successful writers have had similar experiences over the fifty years that the course has been running.

The pandemic that prompted a national lockdown through much of 2020 coincided with the time my group all had to retreat to a quiet place to write our dissertations. Mine was the first three chapters of this book, and lockdown created the freedom from distraction I needed to really concentrate on my writing. I passed the MA with distinction, and more importantly the theme and style of this book had emerged onto paper.

I have intentionally not made more than the odd passing reference to the pandemic and how it has forced us all to challenge some of the assumptions about how we lead our lives. But I am left with a strong sense that history will show that this was the perfect time to write a book that explores a way of life that became lost in the second half of the twentieth century, and I think will slowly be rediscovered in the first half of the present one. I met many people who were challenging recent convention and farming less intensively, choosing to maintain profitability by adding value to what they produced, rather than simply selling it to a wholesaler. Shops and pubs are being saved from closure by collective community ownership, and traditional trades such as blacksmithing are returning, albeit with a contemporary spin.

There is a wonderful quote attributed to the late Douglas Adams that I feel perfectly describes how, by a series of childhood coincidences and a lifelong interest in the work of George Ewart Evans, I came to write the book when I did and, moreover, returned in 2022 to live in Leiston, the Suffolk town where I grew up and where this book begins. Adams summed it up like this: 'I may not have gone where I intended to go, but I think I have ended up where I needed to be.' I hope that, having read this book, you will agree.

Acknowledgements

It is traditional for authors to write a few words of thanks, acknowledging those who have made their book possible. For me, this book is the realisation of an ambition that began to develop when I first read George Ewart Evans's book *Ask the Fellows Who Cut the Hay* in 1969. Sadly many of those I worked alongside on farms back then are no longer around to read this book. Perhaps the greatest influence on my growing interest in rural life has been my father-in-law Michael Easy, to whom I dedicate this book.

It was only when I stopped chasing my tail and joined the Biography and Creative Non-Fiction MA course at the University of East Anglia in 2019 that this book really began to take shape. I'm grateful for the support of Helen Smith, Kathryn Hughes and Ian Thompson, who helped me make the leap from an HND Agriculture gained in 1978 to an MA with distinction in 2020. A key element in that MA is peer support, and the gentle but constructive criticism of my group made a real difference as my writing evolved.

Writing a book is only the start of a long journey, and I'd like to thank Emma Shercliffe, who cast an expert eye over the first draft and suggested I include some more voices, which I know has given the book more shape. My good friend and beta-reader Miles Harvey has also been supportive, both in his feedback on the second draft, but also in helping me to keep going when the crowdfunding campaign that has made publication possible hit a plateau, as crowdfunding campaigns always do.

The team at Unbound also deserve a mention, particularly Mathew Clayton, who said yes, Marissa Constantinou my editor, and Richard Collins and Mary Cheshhyre who helped me fine-tune the book to become the volume you are holding in your hand today.

Finally, I must thank the many people who agreed to be interviewed, who showed me their farms, trades or shared their experiences and memories. And I must not forget all those whose names are printed in the back of this book. Without their willingness to pledge their support, you would not be reading this book.

References

Milk

1 George Ewart Evans, *The Voices of the Children*, Pathian, London, 1947, p. 4.
2 George Ewart Evans, *The Strength of the Hills*, Faber, London, 1983, p. 15.
3 George Ewart Evans, *The Pattern Under the Plough*, Faber, London, 1966, p. 158.

Wheat

4 George Ewart Evans, *Ask the Fellows Who Cut the Hay*, Faber, London, 1956, p. 56.
5 *Ibid.*
6 George Ewart Evans, *Where Beards Wag All*, Faber, London, 1970, p. 133.
7 *Ibid.*, p. 57.
8 George Ewart Evans, *The Farm and the Village*, Faber, London, 1969, p. 48.
9 *Ibid.*
10 George Ewart Evans, *The Horse in the Furrow*, Faber, London, 1960, p. 128.
11 *Ibid.*, p. 49.
12 Evans, *The Farm and the Village*, p. 143.
13 *Ibid.*, p. 145.

Wool

14 Evans, *The Strength of the Hills*, p. 157.
15 Evans, *Ask the Fellows Who Cut the Hay*, p. 33.
16 George Ewart Evans, *The Days That We Have Seen*, Faber, London, 1975, p. 45.
17 *Ibid.*
18 *Ibid.*, p. 46.
19 George Ewart Evans, *Spoken History*, Faber, London, 1987, p. 10.
20 *Ibid.*
21 *Ibid.*
22 *Ibid.*
23 *Ibid.*, p. 11.
24 *Ibid.*
25 Evans, *Ask the Fellows Who Cut the Hay*, p. 157.
26 *Ibid.*
27 *Ibid.*, p. 217.

Leather

28 Evans, *The Horse in the Furrow*, p. 21.
29 George Ewart Evans, *The Crooked Scythe*, Faber, London, 1993, p. 6.
30 Evans, *Spoken History*, p. 16.
31 Evans, *The Crooked Scythe*, p. 7.
32 Evans, *Where Beards Wag All*, p. 205.
33 Evans, *The Horse in the Furrow*, p. 206.
34 *Ibid.*, p. 210.
35 Evans, *The Farm and the Village*, p. 124.
36 Evans, *The Crooked Scythe*, p. 30.
37 *Ibid.*, p. 147.
38 Evans, *The Farm and the Village*, p. 109.

Barley

39 Evans, *Ask the Fellows Who Cut the Hay*, p. 55.
40 *Ibid.*, p. 61.
41 *Ibid.*, p. 62.

References

42 Evans, *The Crooked Scythe*, p. 54.
43 *Ibid.*, p. 55.
44 Evans, *Ask the Fellows Who Cut the Hay*, p. 75.
45 *Ibid.*, p. 77.
46 *Ibid.*
47 Evans, *The Horse in the Furrow*, p. 140.
48 *Ibid.*, p. 102.
49 *Ibid.*
50 *Ibid.*
51 E. Balfour, 'Towards a sustainable agriculture', International Federation of Organic Agriculture Movements Conference, Switzerland, 1977.
52 Evans, *The Farm and the Village*, p. 63.
53 Evans, *Ask the Fellows Who Cut the Hay*, p. 93.
54 *Ibid.*, p. 95.

Coins

55 Evans, *The Strength of the Hills*, p. 9.
56 *Ibid.*
57 Evans, *Ask the Fellows Who Cut the Hay*, p. 28.
58 Isabella Beeton, *Mrs Beeton's Book of Household Management*, Ward, Lock & Co., London, 1909, p. 92.
59 Evans, *Ask the Fellows Who Cut the Hay*, p. 103.
60 *Ibid.*
61 P. A. Graham, *The Revival of English Agriculture*, Jarrold & Sons, London, 1899, p. 23.
62 Evans, *The Horse in the Furrow*, p. 143.
63 Evans, *Where Beards Wag All*, p. 237.
64 *Ibid.*, p. 238.
65 *Ibid.*
66 *Ibid.*, p. 239.
67 Evans, *The Days That We Have Seen*, p. 105.
68 *Ibid.*
69 Evans, *Where Beards Wag All*, p. 135.
70 *Ibid.*, p. 251.

71 *Ibid.*, p. 254.
72 *Ibid.*, p. 252.
73 *Ibid.*, p. 103.
74 *Ibid.*, p. 105.

Coal

75 Gareth Williams, *Ask the Fellows Who Cut the Coal*, Keeling Publications, Pontypridd, 2017, p. 32.
76 *Ibid.*
77 Evans, *The Strength of the Hills*, p. 47.
78 *Ibid.*, p. 37.
79 *Ibid.*
80 *Ibid.*, p. 39.
81 *Ibid.*
82 George Ewart Evans, *The Voices of the Children*, Faber, London, 1983, p. 12.
83 Evans, *The Strength of the Hills*, p. 149.
84 *Ibid.*
85 *Ibid.*, p. 145.
86 David Edward, *The House that Britten Built*, Snape Maltings Trading, Snape, 2013, p. 19.
87 *Ibid.*, p. 29.
88 Evans, *Spoken History*, p. 16.
89 *Ibid.*
90 *Ibid.*
91 Evans, *The Crooked Scythe*, p. 179.

Steam

92 Evans, *Spoken History*, p. 89.
93 *Ibid.*, p. 141.
94 *Ibid.*, p. 142.
95 *Ibid.*, p. 145.
96 *Ibid.*, p. 144.
97 *Ibid.*
98 *Ibid.*

Iron

99 Evans, *The Strength of the Hills*, p. 116.
100 *Ibid.*
101 Evans, *The Horse in the Furrow*, p. 204.
102 *Ibid.*
103 Evans, *The Pattern Under the Plough*, p. 168.
104 *Ibid.*, p. 221.
105 *Ibid.*, p. 224.
106 George Ewart Evans, *Horse Power and Magic*, Faber, London, 1979, p. 138.
107 Evans, *The Horse in the Furrow*, p. 150.
108 *Ibid.*
109 Evans, *The Horse in the Furrow*, p. 52.
110 *Ibid.*, p. 53.
111 *Ibid.*, p. 54.
112 *Ibid.*, p. 55.
113 *Ibid.*, p. 56.
114 *Ibid.*
115 Evans, *Horse Power and Magic*, p. 31.
116 *Ibid.*
117 *Ibid.*, p. 32.
118 Evans, *The Horse in the Furrow*, p. 57.
119 Evans, *Horse Power and Magic*, p. 144.
120 *Ibid.*, p. 146.
121 *Ibid.*, p. 150.
122 *Ibid.*

School

123 Evans, *The Strength of the Hills*, p. 7.
124 *Ibid.*, p. 47.
125 *Ibid.*, p. 39.
126 Evans, *Spoken History*, p. xiii.
127 Gareth Williams, *George Ewart Evans*, University of Wales Press, Cardiff, 1991, p. 33.
128 *Ibid.*

129 Evans, *The Strength of the Hills*, p. 51.
130 Evans, *Where Beards Wag All*, p. 203.
131 *Ibid.*
132 *Ibid.*, p. 205.
133 *Ibid.*
134 Alison Kershaw, 'More than half of young people now going to university, figures show', *Independent*, 27 September 2019, www.independent.co.uk/news/education/education-news/university-students-young-people-over-half-first-time-a9122321.html
135 Evans, *Where Beards Wag All*, p. 205.
136 *Ibid.*, p. 67.
137 *Ibid.*, p. 207.
138 Williams, *George Ewart Evans*, p. 57.
139 *Ibid.*, p. 65.
140 *Ibid.*, p. 58.
141 Evans, *The Days That We Have Seen*, p. 15.
142 Evans, *Spoken History*, p. 123.
143 Williams, *George Ewart Evans*, p. 55.
144 Evans, *Ask the Fellows Who Cut the Hay*, p. 179.
145 *Ibid.*
146 *Ibid.*, p. 180.

Church

147 Evans, *Spoken History*, p. 112.
148 *Ibid.*
149 *Ibid.*
150 *Ibid.*, p. 205.
151 *Ibid.*
152 George Ewart Evans, *From the Mouths of Men*, Faber, London, 1976, p. 72.
153 *Time* magazine, New York City, 21 August 1950.
154 Evans, *The Strength of the Hills*, p. 166.
155 *Ibid.*, p. 167.
156 *Ibid.*, p. 80.
157 *Ibid.*

158 *Ibid.*
159 'A Brief History', The Great Hospital, www.greathospital.org.uk/history/
160 Evans, *Ask the Fellows Who Cut the Hay*, p. 152.
161 Evans, *The Strength of the Hills*, p. 121.
162 *Ibid.*, p. 124.

Trade

163 David Maddox, *Our Story of George Ewart Evans*, Wales, 2017, p. 36.
164 Evans, *The Crooked Scythe*, p. 36.
165 Evans, *The Farm and the Village*, p. 134.
166 *Ibid.*
167 Evans, *The Horse in the Furrow*, p. 223.
168 *Ibid.*
169 Evans, *Where Beards Wag All*, p. 82.
170 Evans, *The Farm and the Village*, p. 128.
171 *Ibid.*
172 Evans, *The Days That We Have Seen*, p. 206.
173 *Ibid.*
174 *Ibid.*
175 *Ibid.*, p. 210.
176 *Ibid.*, p. 211.
177 *Ibid.*, p. 210.
178 Evans, *Horse Power and Magic*, p. 87.
179 Somerleyton Hall & Gardens, www.somerleyton.co.uk/blog-post/somerleyton-estate-trying-to-make-rewilding-in-east-anglia-a-reality/
180 *Sunday Times* magazine, London, 13 June 2021.

Unbound is the world's first crowdfunding publisher, established in 2011.

We believe that wonderful things can happen when you clear a path for people who share a passion. That's why we've built a platform that brings together readers and authors to crowdfund books they believe in – and give fresh ideas that don't fit the traditional mould the chance they deserve.

This book is in your hands because readers made it possible. Everyone who pledged their support is listed below. Join them by visiting unbound.com and supporting a book today.

Ahbeneas
Sarah Alhamad
Nick Allen
Steve Allen
Gareth Armstrong
Sabrina Artus
Martin Ashburn
Adrian Ashton
Neil Astley
Robert M Atwater
Marion Bailey
David Baillie
Tony Barber
M M Barclay

Rodney Barrett
David Barton
Linda Beamish
Alex Begg
Victoria Belcher
Martin Bellamy
Toni Berry
Brad Bigelow
Ali Imran bin Ali Sabri
Adrian Bleese
Graham Blenkin
Alexandra Boliver-Brown
Bob Bones
Alexander Borg

David Boyle
Adam Briggs
Simon Broad
Deborah Brower
Portia Brown
Robert Browne
John Burroughes
John Carlile
Linda Carty
Sam Castriotta
Bridget Chadwick
Lorraine Chamen
Jac Chandross
Thalia Charles
Cherry Tree Farm
Brian Chester
Annie Cholewa
Chris Claxton
Maurice Cohen
Stevyn Colgan
Owen Collins
Chessum Communication
Alan Cudmore
Gill Cummings
Nick Curnock
Nigel Davies
Kathleen Davis
Laura Davis
Jan P. de Jonge

Liz Dexter
Paul Dickson
Ben Doran
Tanza Dryden
Paul Dunning
Frank Eliel
Tom Elliott
Martin Empson
Richard Ferneley
Simon Filbrun
Jane Finlay
Adrian Fisher
Lucas Fothergill
Steven Foyster
Anne Francis
Susan Gardiner
Graham Garner
Steve Garner
Oscar Gennissen
Nicholas Gilbert
Julie Giles
Barbara Gittes
Catherine Glover
Stephen Green
Keith Gregory
Maggie Grenham ^ Rodney
    West
Chris Gribble
Mike Griffiths

Supporters

James Grime
Richard Grove
Andy Gunton
Martin Haggerty
Caroline Hale
Janet Hale
Laurence Hall
Jerry Harrall
Harry Harris
Neil Harrison
Martin Harrop
Miles Harvey
Robin Hawes
Steve Hebson & Fleur Ashworth
Cecilia Hewett
Jeremy Hill
Timothy Hirst
Paul Hodgkin
Xenia Horne
Andy Horton
Susan Housley
Jo Howard
Jackie Howard-Birt
Louise Howlett
Stuart Huddart
Rich Hughes
Richard Hull
Saul Humphrey

Ian F Hunter
Lydia Hutchinson
Paul Jabore
Lee Jaschok
Jane Jeans
Kim Jennings
Alex Johnson
Mary Jordan-Smith
Lizzie Kathiravel
Jo Keeley
Colin Kiddell
Dan Kieran
Kristin Kilgour
Stephen Kimminau
Anne Kinderlerer
Peter King
Jackie Kirkham
Steven Kitchen
Michael Knight
Roman Krznaric
Susan Lansdell
David Lars Chamberlain
Dianne Leafe
Caroline Lee
Chris Liles
Roger Lintott
Clare Lomas
Jane Loveday
David Maddox

Zenda Madge
Dominique Mann
Sarah Marshall
Lucy May Maxwell
Emily Maycock
Stuart Mayhew
Cynthia McKenzie
Colin WD McLean
Cornelia Mews
Andrew Michael
Jason Middleton
Chris Mills
John Mitchinson
Peter Mitson
Margaret Moss
Andrew Mounsey
Rebecca Moyle
Michael Mullan-Jensen
Malcolm Munday
Neville Nancliff
Edward Nash
Carlo Navato
Matt Needham
David Neill
Carol Norton
Angela Osborne
Emma Outten
Frank Paice
Clarissa Palmer
Gwen Papp

Janet Patterson
Mike Pennell
Penny Pepper
Chris Perry
Shelley W. Peterson
Karen F. Pierce
Malcolm Pim
Justin Pollard
Mike Pugh
Caroline Pulver
Nicky Quint
J Avril Radford
Amanda Ramsay
Judy Randon
Kathleen Read
Chris Reeve
Alistair Renwick
Amanda Reynolds
Tara Riddle
Elizabeth Roberts
David Robertson
Arthur Rope
Rosedale Funeral
  Home
Elizabeth Rowlands
Yvonne Rowse
Georgina Rowson
Tim Salmon
Mark Samuelson
Bella Sandcraft

Supporters

Dick Selwood
Daniel Sewell
Sue Sharpe
Victoria Sharratt McConnell
Mike Shaw
Patricia Sheath
Emma Shercliff
Martyn Simonds
Rebecca Sissons
Joanne Smith
Will Smith
Paul Sparshott
Alex Stewart
Paul Stoddart
Kathryn Streatfield
Seija Tattersall
Emma Thimbleby
Cyril Thomas
Liz Thompson
Rosy Thornton
Marian Thorpe
Adam Tinworth
Lesley Townson
Lindsay Trevarthen
Candace Uhlmeyer
Fabio van den Ende
Iain van der Ree
Marian Vincent
Kate Viscardi
Paula Wakefield
Sir Harold Walker
Elizabeth Walne
Jeremy Walton
E Webb
Sarah Whiteman
Miranda Whiting
Trina Whittaker
Gareth Williams
Jenny Williams
Richard Wills
Estelle Wolfers
Andy Wood
Paul Woodgate
Nick Woolley
John Woosnam
Derek Wyatt
Denise Yates

A Note on the Author

Robert Ashton lives near the Suffolk coast, in the town where he grew up. He worked on local farms in his teens, studied agriculture at college and spent the first decade of his career selling fertiliser. His interest in George Ewart Evans dates back to his fourteenth birthday, when his parents bought him a copy of *Ask the Fellows Who Cut the Hay*. Already an established business author, Robert graduated from UEA with a Creative Writing MA in 2020. A Quaker, Robert is driven by a strong sense of social justice and has helped establish a number of social enterprises.

@robertashton1

A Note on the Type

The text of this book is set in Adobe Garamond Pro. Released in 1989, it is a digital adaptation of the roman types of Claude Garamond and the italic types of Robert Granjon. It's one of several versions of Garamond. It is believed that Garamond based his font on Bembo, cut in 1495 by Francesco Griffo in collaboration with the Italian printer Aldus Manutuis. Garamond types were first used in printed books in Paris around 1532.